T0275696

Standalone Photovoltaic (PV) Systems for Disaster Relief and Remote Areas

Standalone Photovoltaic (PV) Systems for Disaster Relief and Remote Areas

Salahuddin Qazi
State University of New York Polytechnic Institute,
Utica, NY, United States

ELSEVIER

AMSTERDAM • BOSTON • HEIDELBERG • LONDON
NEW YORK • OXFORD • PARIS • SAN DIEGO
SAN FRANCISCO • SINGAPORE • SYDNEY • TOKYO

Elsevier
Radarweg 29, PO Box 211, 1000 AE Amsterdam, Netherlands
The Boulevard, Langford Lane, Kidlington, Oxford OX5 1GB, United Kingdom
50 Hampshire Street, 5th Floor, Cambridge, MA 02139, United States

Notices
Knowledge and best practice in this field are constantly changing. As new research and experience
broaden our understanding, changes in research methods, professional practices, or medical
treatment may become necessary.

Practitioners and researchers must always rely on their own experience and knowledge in
evaluating and using any information, methods, compounds, or experiments described herein.
In using such information or methods they should be mindful of their own safety and the safety
of others, including parties for whom they have a professional responsibility.

To the fullest extent of the law, neither the Publisher nor the authors, contributors, or editors,
assume any liability for any injury and/or damage to persons or property as a matter of products
liability, negligence or otherwise, or from any use or operation of any methods, products,
instructions, or ideas contained in the material herein.

British Library Cataloguing-in-Publication Data
A catalogue record for this book is available from the British Library

Library of Congress Cataloging-in-Publication Data
A catalog record for this book is available from the Library of Congress

ISBN: 978-0-12-803022-6

For Information on all Elsevier publications
visit our website at https://www.elsevier.com

  Working together
to grow libraries in
developing countries

www.elsevier.com • www.bookaid.org

Publisher: Joe Hayton
Acquisition Editor: Lisa Reading
Editorial Project Manager: Natasha Welford
Production Project Manager: Susan Li
Designer: Matthew Limbert

Typeset by MPS Limited, Chennai, India

Dedication

This book is dedicated to my late parents, brother Aftab, and my siblings.

Contents

Preface

This book addresses the use of photovoltaic (PV) systems to bring electricity for specific disaster relief and for general supply in remote areas of the world. I was inspired to write this book after realizing that the sunlight striking the earth's surface in just 1 hour delivers more energy than the entire world uses in 1 year. Ironically the countries that get maximum sunlight are currently using very little sunlight to generate electricity despite ever-increasing power shortages that harm their economies. Many of these countries also suffer from unreliable electrical grid supply with frequent power outages, leaving the population even more prone to natural disasters. I grew up in one of these countries.

When natural disaster strikes, local electrical power in most cases is the first utility affected. Disruption of electricity can last for weeks as transmission lines are slowly repaired, causing deaths and economic loss. In 2013, the United Nations reported that in the past 20 years approximately 4.4 billion people have been affected by natural disasters that claimed 1.3 million lives and cost the world $2.0 trillion in economic loss. In the aftermath of these disasters, the PV systems can bring natural, reliable power to places recently devastated. PV systems in the standalone mode can also provide electricity for more than 1.3 billion people worldwide who do not have access to grid electricity. This figure includes more than 500 million people in Sub-Saharan Africa, more than 300 million in India, and approximately 2.6 billion people who use pollutant-emitting wood, biomass, or dung for cooking and heating their homes. For increasing the affordability and cutting the costs of rather expensive PV systems, some countries in the developing world are giving incentives through micro-financing institutions and solar companies that couple their solar products with financing plans, based on customer income. In the United States, new partnerships are emerging between citizens, electrical utilities, and governments to deliver electricity for low-income households at a reduced cost. In 2015, the White House announced that 68 cities, States, and businesses have signed up for an initiative to promote community solar with an emphasis on low- to moderate-income households.

Global installation of solar PV is expected to reach 64.7 GW in 2016, with the largest market being China at 19.5 GW capacity, followed by the United States and Japan. The new international solar energy alliance of 120 countries announced at the Paris COP21 Climate Summit indicates the importance of solar energy, its application to clean energy, and poverty alleviation for millions of people living in the developing world with no access to the grid. Under this alliance, India plans to install 175 GW of renewable energy by 2022 to provide electricity to each Indian household. In the Middle East, Dubai announced a large program of mandating to install solar panels on all rooftops by 2030 as part of a plan to make the city a global clean energy center.

This book of mine is an introduction to various PV systems that can meet the immediate needs of heating, cooling, and lighting homes and businesses in the aftermath of a disaster. The book also shows ways that PV systems can deliver electricity to remote areas for everyday living including pumping water, purifying water, powering medical clinics, powering cell phones, powering mobile communication base stations and powering other facilities for off-grid areas. The book contains less mathematical derivations, cites more case studies, examples of practical PV systems, figures, software tools, databases and appropriate websites to supplement the chapter topics. This hands-on approach makes the book a must-read for designers, installers, students of renewable energy programs, policy makers, non-governmental organizations and all other people working in disaster recovery programs. It serves both the technical and non-technical readers. The book cites the names of many organizations and corporations that have developed products for PV systems, software tools, databases, and websites. The mention of a specific company name or product must not be misunderstood as an endorsement of any products nor suggest that any product is superior to another.

Chapter 1, Photovoltaics for Disaster Relief and Remote Areas, discusses differing types of natural disasters, their effect on peoples' lives and on the world economy, their impact on electrical grids and the causes and costs of power outages. Chapter 2, Fundamentals of Standalone Photovoltaic Systems, presents the principles, types, and components of standalone PV systems including solar sources, PV cells, PV modules, PV arrays, inverters, charge controllers, and types of storage batteries as well as the design of standalone systems, sizing methodologies, and details of modeling software for performance analysis and simulation of PV systems. Chapter 3, Mobile Photovoltaic Systems for Disaster Relief and Remote Areas, Chapter 4, Portable Standalone PV Systems for Disaster Relief and Remote Areas, and Chapter 5, Fixed Standalone PV Systems for Disaster Relief and Remote Areas, discusses three standalone PV systems and their benefits, methods of deployment, examples of practical systems and case studies related to appropriate topics for each chapter. The case studies include review and detailed analysis of select PV systems that can be applied for disaster relief and remote areas. Chapter 6, PV Systems Affordability, Community Solar, and Solar Microgrids, is related to affordable PV systems, community solar, and solar micro-grids. This chapter discusses new initiatives and programs for providing solar-generated electricity to low-income people both in the United States and the developing world at low cost as well as the ways of making the electric grid more resilience to outages. Chapter 6 also discusses two case studies about solar micro-grids in the United States and India.

Chapter 7, Solar Thermal Electricity and Solar Insolation, is divided in two parts. The first part is on the working principles and types of concentrated solar power technologies used in the United States and the rest of the world. The second part describes solar insolation/radiation, and the methods of measuring them using ground measurement and satellite images. It also provides a list of online databases for estimation of solar insolation/radiation.

The book also has a comprehensive appendix on the following topics:

- Building dye-sensitized solar cells
- Assembling one's own PV module
- Results of solar insolation/radiation using different databases
- Standards for PV systems.

Salahuddin Qazi
Professor Emeritus, State University of New York Polytechnic
Institute, Utica, NY, United States

Acknowledgments

I could write this book only with the encouragement and support of my wife, Dr. Naseem Ishaq and my son Farhan A. Qazi, who is a full-time employee and doctoral candidate student. My wife reviewed and corrected my manuscript, and gave technical advice and constructive criticism. My son helped me with technology support and presentation. I am grateful to them.

I thank Dr. Richard Komp, Director of Skyheat Associates, for invaluable suggestions. Thanks are also due to Mr. William Young of SunTree Consulting for discussing the initial outline of the book and writing a joint paper on "Disaster relief management and resilience using photovoltaic energy."

I would like to acknowledge the critical role of Joseph Hayton, Lisa Reading, and Natasha Welford of Elsevier in starting the book. Natasha supported me throughout the duration of writing the book with necessary information and advice. Thanks are also due to all of the corporations and organizations that have allowed me to republish their product images.

Chapter 1

Photovoltaics for Disaster Relief and Remote Areas

1.1 INTRODUCTION

Disasters are part of everyday life, impacting every region of the United States, and the world, in increasing number and severity. These often unpredictable events not only shatter the lives of a few people very quickly, but cause lingering problems in the aftermath for many. When a serious disaster strikes, local electrical power in most cases is the first utility to be affected, at least temporarily. Disruption of electricity can last for weeks on end as transmission lines are repaired, causing deaths from failed medical equipment, heat or cold, and lack of amenities for sustaining everyday living. For assisted living facilities and nursing homes, life safety is at risk and is critical to maintain. Many businesses and financial institutions suffer economic losses from the interruption and downtime of highly sensitive loads such as data centers and related services. Disruption of electricity also affects malfunctioning of gas stations, cell towers, and emergency call centers with far-reaching social impacts. In populated areas, after a small disaster, sagging electrical lines can affect large regions by triggering burning of utility poles, creating power outages and disrupting power to millions of people by a chain reaction.

In the aftermath of any disaster, photovoltaics (PV) has the potential to help bring natural, reliable power to places devastated by these events. The use of diesel-powered engines commonly used to provide emergency power can unfortunately be dangerous in the hands of untrained users and is reported to result in incidents of fire, fuel explosion, burns and problems of noise. Generating electricity from solar energy through the process of photovoltaics can be used to heat, cool, and light our homes and businesses. Photovoltaic power systems provide emission-free electricity fueled by the sun which is reliable, secure, noise free, and does not need refueling. It also helps to reduce consumption of fossil fuels in power plants, decreasing pollution and greenhouse gas emissions that cause climate change.

Photovoltaic systems in an off-grid mode can also be used for more than a billion people living in remote areas with no access to electricity. According

Standalone Photovoltaic (PV) Systems for Disaster Relief and Remote Areas.
DOI: http://dx.doi.org/10.1016/B978-0-12-803022-6.00001-0

to one statistic, roughly one out of every four people (about 1.6 billion) in the world do not have regular access to electricity. Because of the lack of electricity, billions of people use wood, biomass, or dung for cooking and heating their homes. Such traditional methods of generating energy results in serious environmental and health problems, including massive deforestation, sickness, and death. Photovoltaics can generate clean and sustainable electricity in remote areas that can improve health, provide cleaner environment and help in boosting agriculture, starting new businesses, and creating jobs that lead to greener economic development.

Energy demand continues to increase with the need for new and clean energy sources. Solar photovoltaic will definitely become a vital source of energy, particularly for meeting demand after disasters and in remote areas. On the industrial and business side, the price of PV systems has decreased by more than 50% and the global PV market has grown at a compound annual growth rate of about 40% for the past 10 years. In 2013, PV developers installed 37.5 GW of new panels, increasing to 38.4 GW in 2014. The cumulative worldwide capacity increased to 138.9 GW in 2013 despite the recent economic crisis. In 2014, the United States installed 1330 MW of solar PV, which was 79% more than 2013. There are now over 175,000 MW of cumulative solar electricity capacity with nearly 600,000 individual online systems; this sum is expected to reach 1,000,000 in 2015. China surpassed Germany to become the world's largest solar market in 2013, and is predicted to install approximately 19.5 GW in 2016.

According to the NPD Solarbuzz Marketbuzz report, the global solar PV industry is headed to a 5 year growth in the cumulative installed capacity of 500 GW by 2018. The report also predicts 100 GW of solar deployment in 2018 with annual projected revenue of $50 billion.

1.2   TYPE OF NATURAL DISASTERS

A disaster can be defined as a sudden, calamitous event that disrupts the functioning of a community and causes economic, environmental, human and material losses that exceeds the community's ability to cope using its own resources. In term of human losses, an event is categorized as natural disaster if more than 10 people are killed or at least 100 people are injured, displaced, evacuated, and made homeless. The event is also classified as a natural disaster, if a particular country declares it as such or if a country declares it a natural disaster and asks for international assistance. The disasters can be divided in two broad categories: natural disasters and man-made disasters. Natural disasters include avalanches, earthquakes, forest fires, floods, hurricanes, landslides, tornadoes and others. Man-made disasters include airplane crashes, fires, nuclear power plant accidents, terrorism, and wars. Both categories of disasters are unexpected, sudden, and uncontrollable. The effect of natural disasters lasts much longer and is accompanied by disruptions to critical infrastructure systems including electrical power systems, transportation, and water. Natural disasters are naturally-occurring physical phenomena caused either by rapid or slow onset events which can include the following:

- Geophysical: Originating from solid earth and includes earthquakes, tsunamis, volcanoes, landslides, dry rock falls, avalanche.
- Metrological: caused by short-lived/small to meso scale atmospheric processes (in the spectrum from minutes to days). This includes storms, tropical cyclones (hurricane), snow storms, heat/cold waves, drought and wildfires.
- Hydrological: caused by deviations in normal water cycle and/or overflow of bodies of water caused by wind set up such as floods and avalanches.
- Climatological: caused by long-lived/meso to macro scale processes (in the spectrum from intraseasonal to multidecadal climate variability). This includes extreme temperatures, droughts and wildfires.
- Biological: caused by the exposure of living organisms to germs and toxic substances. Include disease epidemic, insect/animal plagues.

Climate-related disasters include metrological events—these have greatly increased while the number of geophysical disasters has remained fairly stable since the 1970s. Tropical storms or hurricanes are among the most powerful natural disasters by virtue of their size and destructive potential. Tornadoes, on the other hand, are relatively short-lived but violent and can cause winds more than 200 miles/hour. Flooding is the most common natural hazard while tornadoes and earthquakes strike suddenly without warning. Wildfires are more prevalent in areas experiencing drought.

According to recent reports, the number of natural disasters worldwide has steadily increased since 1970. It is also reported that the number of natural disasters is highest in North America such as hurricanes, floods, snow storms, severe heat, tornadoes, and even drought. There were three times more natural disasters between 2000 and 2009 than between 1980 and 1989. The number of people affected by natural disaster have increased to 217 million every year since 1990 and almost 300 million people now live around the world with insecurity. The *New England Journal of Medicine* has reported that the scale of disasters has expanded because of increased rates of urbanization, deforestation, and environmental degradation. In addition, intensifying climate variables such as higher temperatures, extreme precipitation, and more violent wind/water storms contribute to more disasters.

When a disaster strikes, the whole infrastructure, including electricity shuts down for days or longer depending on the nature of the disaster. In the absence of electricity, all human activities and businesses are either damaged or ruined. According to a recent report by the World Bank, natural disasters have cost the world US$3.8 trillion during the 1980–2012 period. It is also estimated that the annual cost from these damages has increased from US$50 billion a year in 1980 to US$200 billion a year in the last decade including the era of Hurricane Katrina (2015) and Superstorm Sandy (2012) in the United States. The cost of natural disasters in 3 of the last 4 years in the last decade exceeded US$200 million. The majority of these losses (almost 75%) are related to extreme climate events.

Weather-related disasters affect people the same way whether they occur in developed or developing countries. The poor, vulnerable, and less-prepared nations suffers the most from the disasters. According to an estimate, 70% of the world's disaster hot spots are located in low income countries and 33% of world's poor live in multihazard zones which makes them highly vulnerable to disasters. As a result, low-income countries have accounted for 48% of fatalities with only 9% of disaster events since 1980. This can have a devastating economic impact for the developing countries. The World Bank Group and Global Facility for Disaster Reduction and Recovery analysis shows that the disaster's impact on gross domestic product is 20 times higher in developing countries than in developed countries. As global climate continues to change, developing countries are more vulnerable to the losses from increased activity of disasters such as floods, storms, and drought. It is estimated that there could be 325 million people vulnerable to weather-related events and prone to poverty in sub-Saharan and South Asian regions by 2030. Many large coastal cities in the growing middle-income countries could face combined annual losses approaching $1 trillion from weather-related disasters by the middle of the century.

Natural disasters can cause severe damage to electrical power systems and disrupt electricity for distribution to the customers in affected areas. The damage occurs as a result of vibration, fire, or floodwater due to hurricanes, storms, and tornadoes packed with high winds which can lead to the malfunction of electrical equipment and power lines.

1.3 ELECTRICAL POWER SYSTEM/GRID

An electric power system is defined as a network of electrical components used to supply (generate), transmit, and consume electric power. An electric power system that supplies power to homes and industries for a sizeable region is called an electric grid. Electric grids can be divided into three-layered complex interconnected networks consisting of generation, transmission, and distribution components. An electric grid also contains control software and associated equipment to transmit electricity from the place of generation to residential, industrial, or commercial users. This is achieved by transporting electrical power from generation buses to distribution substations through transmission buses interconnected by transmission lines. The point of generation is usually located in a centralized point which is far removed from the place where it is consumed. Electrical energy is generated in the power plant by transforming other sources of energy. These sources include chemical, heat, hydraulic, mechanical, geothermal, nuclear, solar, and wind which can be used to produce electrical energy. Electrical energy produced from this conversion is then transformed to high voltage which is more suitable for efficient long distance transportation to the consumption locations using high voltage power lines. The high voltage electrical energy is stepped down to lower voltage by transformers at the substations so that it is suitable to distribute for residential, commercial and industrial consumption. The end user is connected to lower voltages which are obtained by a stepdown process using a transformer in several stages as network is reduced in capacity. The grid can be connected to a single power source or electricity generating plant but is usually linked with other plants to provide a more flexible and reliable network.

Depending on the national standard, typical voltage for domestic consumer is 120 or 220 V single-phase alternating current (AC). The majority of the power systems used to distribute and supply electricity directly to higher power equipment is three-phase AC which is the standard across the world. Power at the smaller scale is generated by smaller systems which are often used for hospitals, universities, industrial units, and commercial buildings. For factory and workshop machinery using high power equipment, 210 or 415 V three-phase electricity is used.

The US electric grid generates, transmits, and distributes electric power to 144 million end-user customers by delivering to homes, offices, schools, factories across the country. The grid consist of high voltage transmission lines, local distribution systems, power management, and control systems that connects Americans with 5800 major power stations. This includes over 450,000 miles of high voltage transmission. The total capacity of power generation from major power plants amount to approximately 1000 GW. The current US administration has been actively investing in a 21st century modernized electric grid which will be more efficient, secure, reliable, and resilient to external and internal causes of power outages.

1.3.1   Impact of Disasters on Power Systems/Grid

Electrical power systems are real-time energy delivery systems, which means that the power is generated, transported, and supplied when the power switch is turned on. These systems do not store electrical energy and instead generate electrical power as the demand calls for it. As a rule, the electrical power systems are designed to operate during relatively stable weather conditions and proper loading conditions. However, these design assumptions may be strained due to extreme weather. The most vulnerable electrical equipment to be affected by the disaster damage is high-voltage transformers located within and outside the substations. These transformers in the substations are an essential part of transmission lines and are large, heavy and difficult to move. Most of these devices are custom built, hence difficult to replace as the delivery might take a long time. Other vulnerabilities from natural disasters include threats to generators, transmission lines, substations and control centers. Natural disasters can disrupt fuel supply and can also damage transmission lines. The loss from disasters affecting a control center which coordinates the operation of the grid to maintain reliability can have serious impact on the operation of the grid.

Large centralized power stations including critical systems such as control software, communications, and sensors are vulnerable to man-made cyber-attacks through Internet connections or by direct incursion at remote sites. Any telecommunication link that is even partially outside the control of the system operators could be an insecure pathway into operations and a threat to the grid. Transmission lines and substations with high voltage transformers are particularly vulnerable to terrorist attacks. Transmission lines can be replaced easily but transformers takes a long time to replace.

Power lines can be downed or sagged by falling trees as a result of strong winds during hurricanes and storms. A sagging power line shorting out by a falling tree or vegetation can initiate cascading effects resulting into power outages. In Ohio in 2003, a blackout originated with the FirstEnergy power plants and transmission lines. The Northeastern blackout of 2003 originated in Ohio with FirstEnergy Corp which led to the loss of power for more than 50 million people in United States and Canada. The blackout started with the FirstEnergy power plants and transmission lines when three of its four lines carrying 345 kV sagged and touched the fallen trees and disconnected these lines by tripping their circuit breaker. In addition the alarm system that was intended to notify the utility about the failure of lines, also failed. The combined effect of these two failures resulted in a large drop in voltage across the entire power network. As a result the low voltage and high current on the nonfailed lines further tripped the circuit breaker on another 345 KV line (Sammis-Star), setting off the cascade. This kind of situation can be avoided either by burying power lines or cutting the vegetation and trees around power lines. According to the Energy Information Administration, constructing underground lines per mile is 5–10 times more expensive (US$500,000 to US$2 million per mile) than installing overhead lines. It costs more to repair underground lines as they do not last very long and need to be dug up when they get old or are broken. Underground lines are more vulnerable to floods, while overhead lines are more vulnerable to storms. In the United States only 0.5% of long distance transmission lines are underground and 18% of distribution lines are underground. This obviously leads to spending less money on cutting the trees to protect the overhead lines but still making it more vulnerable to storms and tornadoes.

1.4 CAUSES OF POWER OUTAGES

A power outage or blackout is defined as power interruption during which a customer has lost access to electricity grid. It is measured in duration of time. In United States, the time for such an interruption of power is less than 5 minutes. In United Kingdom, this interruption is defined as more than 3 minutes and in Sweden it is 1 minute. A large power outage is normally referred to as a blackout affecting more than 1 million people. A recent example of large power outages include two severe blackouts in India in Jul. 30 and 31, 2012, where more than 670 million people, half of India population spread across 22 states in Northern,

Eastern and Northeast India were affected. The power outage took off an estimated 32 GW of electricity.

Either types of natural disaster could damage many electric power system components, causing widespread outages or power shortages over a short duration or a longer period of time. The damages leading to longer time outages can happen only if significant components of generation and transmission are damaged. The most probable causes of such damage are hurricanes or strong earthquakes which affect distribution systems much more than generation and transmission.

The recent example of a hurricane is Superstorm Sandy which made landfall in southern New Jersey, United States, on the evening of Oct. 29, 2012. Superstorm Sandy affected 24 states, including the entire eastern seaboard from Florida to Maine, and west from Michigan to Wisconsin with winds spanning 1100 miles/hour. The estimated damage totaled over US$68 billion with fatalities of at least 181 people. More than 8.5 million houses and businesses experienced power outages which took several weeks to be restored in certain areas. One quarter of cell sites in 10 states went down and ground transport was disrupted for days.

The other extreme weather events responsible for long term power outages is floodwater from heavy rain and storms which can damage the underground electrical systems and power substations. Because of large amount of water, rust and left over trapped mud, it takes much longer to restore a flooded substation than restoring a downed power line damaged by wind or ice.

Massive storms and hurricanes inflict damage in two different ways. First, the gushing flow of water will flood the low-lying coastal areas, damaging the buildings and houses. The second is the damage to the power generating equipment in the way of water, which is of great concern to nuclear plants in the flood's projected path. In case of Japan's Fukushima nuclear meltdown, it was the tsunami's surge of seawater that knocked out the electrical generators and not the earthquake itself. The Fukushima nuclear disaster took place on Mar. 11, 2011 at Fukushima I nuclear power plant and resulted in a meltdown of three of plant's six nuclear reactors. The meltdown was the result of 33-feet tsunami triggered by earthquake of magnitude 9.0. Flooding from the tsunami destroyed backup generators for the Fukushima I nuclear plant, resulting in lost power for 4.4 million households. According to one estimate, it could take years before full power generation is restored to pre-earthquake levels. The electrical components in the power substations, such as pumps, relay panels, transformer fans, switchgears are the most vulnerable pieces of equipment, and need to be dried once all the water has been pumped out. Other items such as cylinders, bearing, pins, rings, gaskets, latches may need to be replaced and bearing breaker mechanisms may have to take apart and cleaned manually.

The persistent heat waves in certain part of the world can affect transmission lines' sag and other components such as transformers and rotating machinery that needs to be cooled off. The sagging lines are vulnerable to falling trees, flying debris, and flooding. Indeed, some types of transmission towers cannot withstand high winds. According to recent studies, the US economy has suffered costs of between US$20 billion–US$55 billion per year due to storm-related power outages. Other causes of weather related power outages are given below:

- *Wind, tornadoes, and hurricanes*
  High winds or fallen tree due to tornadoes or hurricanes may cause a transmission line to touch and short circuit, causing the circuit breaker to trip and produce an a outage. The tree or its limbs may be blown onto the power lines by wind and make the lines fall to the ground, breaking the lines and poles.
- *Rain and flooding*
  Flooding as result of heavy rains can cause damage to both aboveground and underground lines and equipment. The utility may shut down the electricity in order to prevent damage to the equipment, affecting service to some customers.
- *Snow and ice*
  Snow and ice buildup during a storm can increase the weight of ice on a tree and its branches, causing it to fall onto power lines. This will in turn knock down the power lines and poles or knock down the lines onto each other, causing an outage.
- *Electric lightning*
  The occurrence and striking of lightning after a storm can cause outages if it hits certain electrical equipment. It can also strike a tree, causing it to fall onto power lines and resulting in an outage.
- *Cascading effect*
  Large power outages are the result of a small and single event that gradually leads to cascading outages and eventually collapsing the entire system. Large power outages are more complex as it involves cascading failure in several power systems and affects a large area of grid. A cascading outage is a sequence of failures and automatic disconnections of transmission lines and power generators as a result of an initiating event due to natural causes or human actions. Natural causes include sagging of line in to vegetation or falling tree due to high winds or shorting of line by lightning. The human actions or inactions include lack of monitoring or overheating of lines due to imbalance between power generation and load, resulting in overheating and sagging of power lines. Recent cascading failures include some of the following power outages. In United States the worst blackout happened in the United States-Canada region on Aug. 2003, affecting eight states and two provinces in Canada and left 50 million people without power. The combined load of 61.8 GW was lost for up to 4 days. Other power outages include the Italian blackout of Sep. 28, 2003 which affected 57 million people and around 19,000 MW of electricity load was lost over a 277,000 km^2 area. The Scandinavian blackout of Sep. 23, 2003 affected approximately 5 million people, cutting off around 3000 MW of generating capacity in Sweden, and 1850 MW in Denmark.

1.4.1   Cost of Power Outages

The cost of outages take various forms such as damage to business and property, lost output at the factories, spoiled inventory, delayed production, loss of livelihood, and damage to the electricity grid and its associated equipment. According to a recent report weather and climate-related disasters have caused the world more than US$2.4 trillion economic losses and nearly 2 million deaths since 1971. The report focusing on six types of disasters: droughts, floods, extreme temperature, storms, wildfires, and landslides from 1971 to 2010, indicate that each disaster frequency has risen and climate continue to change. The number of natural disasters from 2000 to 2009 has increased three times than the years between 1980 and 1989 with 80% of this growth due to climate-related events.

Asia, over the last 20 years, according to Asian Development, accounts for half of the world's estimated economic losses from extreme weather events. It is also estimated that the region will experience US$53.8 billion in losses annually from disasters which are predicted to increase both in severity of extreme weather and frequency. Businesses, citizens and governments will be facing both expected and unexpected challenges including scarcity of resources, rising cost and long-term impact on the environment due to extreme weather conditions.

In the United States, the cost of outages was analyzed in 2013 by a White House report, which indicated that there were 679 weather-related power outages between 2003 and 2012. These outages cost the US economy an inflation adjusted annual average of $18 billion to $33 billion. This report entitled "Economic Benefits of Increasing Electric Grid Resilience to Weather Outages" also indicated that each outage affected at least 50,000 customers. Annual costs for each weather-related disaster fluctuate significantly and are the greatest in the years that major storms occur. Hurricane Ike in 2008 cost approximately US$40 billion to US$75 billion, Hurricane Irene in 2011 cost an estimated US$15.8 billion in total damages, and Superstorm Sandy in 2012 cost an estimated US$27 billion to US$52 billion. This variation in estimates reflects different assumptions and data used in the estimation process.

It is recommended that continued and timely investment in grid modernization and resilience will diminish these costs over time. This strategy, coupled with regular inspections and timely maintenance of existing grid infrastructure, will save the national economy billions of dollars and reducing the hardship experienced by millions of people by the weather-related disasters.

1.5   ENERGY NEEDS IN THE AFTERMATH OF DISASTERS

The major challenges faced after every disaster include:

- Pre-disaster early warning infrastructure
- Supply of food and clean drinking water
- Health and sanitation
- Information and communication

- Power and energy for lighting and cooking
- Waste collection and disposal, including rapid disposal of dead bodies of humans and animals
- Disaster-proof housing and shelter
- Emergency and post-disaster shelters
- Rescue and relief operations
- Transport infrastructure

These needs are grouped into five general categories as explained below.

1.5.1 Backup Power for Emergency Shelters

Backup power is urgently needed during emergencies for medical services, clean water, emergency lights, communications and electrical services, among other needs. In hospitals, patients' lives could be at risk if power is not available for operating rooms, medicine refrigeration, and life support systems. The backup power needed to sustain life in hospital is more than 3 days which is typically needed for certain types of critical equipment after a disaster. The use of diesel power generators is one way of providing backup power which is usually placed on the roof. These power sources, however are heavy, difficult to renew, and bad for the environment. Photovoltaic-based mobile and fixed generators can be used to generate electricity for emergency shelters and provide backup power to supplement other energy sources. Examples of such applications include solar installation for gas stations combined with battery storage to operate fuel pumps in the aftermath of disasters. Supply of clean water and water purification can also be achieved with solar-powered water pumps and purification systems applied to a well connected to groundwater source for emergency shelters.

Solar powered shelters have been installed in different parts of the United States and the world. In the state of Florida, United States, more than 115 10 kW PV systems for emergency shelter schools have been installed under Smart E-Shelter Program with the coordination of UCF's Florida Solar Energy Center (FSEC). FSEC started this program in 2003 after participating in a US Department of Energy program to raise awareness and understanding of photovoltaic technology among students, teachers, and the general public. In 2010, these PV systems were moved from a demonstration project to a viable application level. The reason for this move was a power outage in Florida, for which FSEC was awarded a contract by the DOE administered American Recovery and Reinvestment Act to install PV systems at shelter-designated schools. Since then, the Federal Emergency Management Agency (FEMA) has revised its Standards to make shelters more disaster-resistant and comfortable for those displaced by disasters. Each PV system is ground mounted and consist of 10 kW solar array, an inverter/charger and 25 kWh battery for backup power. These PV systems are used for powering items such as phone chargers, radios, overhead lights, and life support equipment using both DC/AC voltages during and after the disaster. It is important for schools, businesses, and factories to benefit from keeping a solar powered disaster kit consisting of radio, phone charger, battery charger flashlight, and other devices.

IKEA and UNCHR have recently developed a solar powered shelter for displaced citizens in different parts of the world. This solar-powered shelter is a 188-square-foot hut with solar panel roofing allows users to generate their own electricity. These shelters come flat packed, easy to transport and can be assembled in 4 hours. The hut can accommodate five people to sleep and can last for 3 years. The solar panel roofing deflects 70% of sun rays, making the interior cooler during the day and warmer at night. This shelter is more comfortable than a tent and is ideal for more than 40 million refugees worldwide who are displaced from their homes.

1.5.2   Emergency Lighting

Emergency lighting is vital not only for ensuring the safe and easy navigation and evacuation of properties, but also for street lights, traffic lights, and portable personal lights during unexpected events. Solar-powered lights, both portable and fixed, can be used for disaster-affected areas. Fixed solar powered street lights are usually lower in power than the traditional street lights but are equally important to illuminate the dark areas in the aftermath of disasters. These solar powered lights can also be used to illuminate highway signs, parking lots, parks, bus shelters, and other remote areas. Portable solar powered light sources including solar lanterns and inexpensive solar gardens lights can be used to illuminate areas beyond the reach of fixed light sources.

Solar powered light sources are used not only for emergencies but also for remote areas on a continual basis. These light sources in many cases are economical, convenient, and easier to implement than extending existing utility services to remote locations. Most of the portable solar lights use alkaline batteries, which are easily available and can last a long time.

1.5.3   Communications

In the event of a disaster, local communication systems' infrastructure becomes inoperative because of either an electrical failure or destruction of transmission towers. These communication systems include the wireline and wireless telephone networks, broadcast and cable television, radio, public safety land mobile radio, satellite systems, and the Internet. Some of these systems, such as radio broadcasting and television are most effective for sending disaster warnings, both in the developed and developing countries. In the absence of grid electricity in the aftermath of disasters, installing and maintaining diesel generators may not be cost effective compared to solar-based systems which can also provide clean electricity. Solar-powered systems can be used to power some of these communication systems particularly low powers Wi-Fi wireless networks which can be used with communication repeaters and variable message boards for roadside

IP links, traffic management and other public communications. PV systems are good alternatives to traditional electrical power for remote wireless networks and are cheaper to install than running an electrical power line to the grid.

1.5.4   Transportation

As the delivery and dispensing of gasoline depends on electricity at many stages, many people in the aftermath of disasters are unable to use their gasoline-powered vehicles for extended periods of time. Without the proper transportation at times of emergency, many problems including evacuating people from hazardous areas and providing the necessary supplies to sustain life could be created. Solar-powered vehicles have the potential to become a form of independent transportation for procuring supplies during an extended disaster. This can be achieved by using small solar golf carts or 100% electric cars that employs solar panels to charge the battery. Solar golf carts are low speed vehicles that use solar panels or thin film solar cells to power the electrical motor driving the golf cart. Solar panels made of silicon or thin film are mounted on top of the existing roof or the panels can be used as the roof itself. The batteries in the golf cart are charged by the sun's energy. The University of California San Diego has developed a solar-powered golf cart (shown in Fig. 1.1) that can achieve a speed of 25 miles/hour and can be used for moving things around the disaster area. The golf cart uses a PV roof of about $2\,m^2$ (15% efficiency) to generate 300 W from an average of 5 full-sun hours per day. It uses a battery of 5 kW.

Solar conversion kits ranging from low wattage solar battery chargers to a 410-W array on an 8-passenger transport cart have been available for several years. Thin film flexible solar panels are often preferred on golf carts due to their light weight and ability to conform to the shape of the cart roof.

FIGURE 1.1   Solar-powered golf cart. *Rhett Miller, UC San Diego Resource Management and Planning.*

Electric cars are like solar cars but get their energy from batteries instead of solar panels. The batteries must be recharged when they run out by plugging the car into an electric power outlet. The electric cars such as Active E, Chevy Volt, Mini E, Nissan Leaf, and Tesla are already in production and many major car manufacturers offer some version of plug-in electrical vehicle or hybrid such as Toyota Prius and Ford Fusion. In the absence of grid electricity after the disaster, solar power can be used to charge the batteries in your own garage overnight if your home is solar powered instead of filling these cars up at the gas station.

Solar panels can also be mounted on the roof of cars, but that will provide insufficient power to charge the car battery because of the size of these panels.

In order to boost the intensity of light captured by the solar panels on the roof of a car, Ford Motors Inc., has come up with a new concept based on concentrators. The concentrator is a large Fresnel lens acting as a magnifying lens, placed in an acrylic canopy that stands over the car consisting of PV roof to charge the car's battery. The Ford C-MAX Solar Energi Concept car consist of $1.5\,m^2$ of solar panels on its roof which generate only $300\,W$ of power that is not enough to charge the vehicle's battery in 1 day. However, by coupling the car with a carport that has a concentrating lens atop it, solar energy will be boosted 10 times the solar energy that the vehicle can produce alone. To keep the car under the highest concentration of sunlight, the canopy magnifier is tied to the car's on-board computer which will move the car as the sun moves from east to west. This will allow the car's battery to be recharged in a single day. The C-MAX Solar also have the option to power up via the grid, if necessary.

- *Vehicles for Disaster Relief*

  Vehicles of different sizes equipped with solar panels are also designed to offer immediate portable power in areas hit by disasters. Sedan cars like Toyota Prius with solar roofs have been used to power electronic devices for police, medical services or fire emergency vehicles. Fire trucks have been used to power electronic devices and backup power by installing solar panels, charge controllers, wiring and other accessories. NRG Cars has developed a 42 foot transport vehicle called a Power 2 serve vehicle with a 26-foot trailer with solar panels photovoltaics and generators to travel to disaster-struck regions and help with any immediate needs. The vehicle consists of $10\,kW$ solar panels, and two diesel generators of 10 and $20\,kW$ each as shown in Fig. 1.2.

  The vehicles (truck and trailer combination) can provide power up to three homes, 100 charging stations for cell phones, tablet computer, cameras, power equipment and small tools, flat screen TVs, Wi-Fi and access satellite, internet service and provides emergency services to people in the shelter after disasters. The truck can be deployed as a 50×20-foot pavilion with air conditioning and heating facilities for disaster victims. The system also has a battery for 2 hour storage and sleeping accommodation for up to seven people.

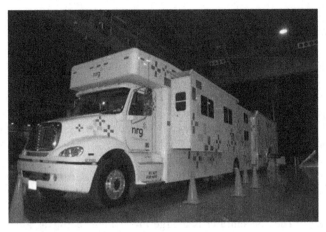

FIGURE 1.2 NRG vehicle for disaster relief. *NRG Energy.*

1.5.5 Portable Systems and Battery Charging for Miscellaneous Applications

Portable solar systems are usually easy to transport, and small in size, hence can be used to provide electricity for disaster relief. Each system consists of solar panels, solar charge controllers, battery, wiring, a carry case or pack, and sources for 12-V DC and/or 120-V AC power. These systems are available in a variety of configurations and energy potential and can be used to provide electricity for relief operation centers, and to operate lights, fans, small radios, small TVs, vaccine refrigerators, and medical equipment. The goal of each system is to provide power for electronic devices during disaster crisis when there is no time for careful installation of delicate equipment. Portable solar systems with storage batteries have been used for relief efforts, construction tools, water purification systems, remote sensing applications, air monitoring and many other applications. In the aftermath of disasters, it is crucial that rechargeable batteries used in various electronic devices both by rescue crew and victims should be charged in a timely manner. Failing to do so can interrupt communication between the rescue workers, victims and can cause hazards for not maintaining essential medical services. For long-term power outages in the case of disasters and remote areas, an appropriate solar charger that charges an external battery is the requirement. Portable solar systems will require portable solar chargers.

- *Solar battery charging*
  Solar chargers employ energy from the sun to supply electricity to power or charge electronic devices. Solar chargers can charge small Ni-Cd rechargeable batteries of 1.2, 3, 6 or 9V of several hundred milliampere

hours. They can also charge large lead acid batteries of 12, 24 or 48V to several hundred amperes hours. Small batteries are often used in portable devices even though some portable devices can be powered directly from the sun without battery backup. Rechargeable batteries are commonly found in cell phones, cordless power tools, cellular and cordless phones, hand-held radios, laptop computers, digital cameras, two-way radios, camcorders, portable medical devices, and remote control toys. Solar powered battery chargers are available to charge AA, AAA, C and D type batteries. Solar chargers on wheels are also available which can be transported from one place to another and can be used by many people.

One example of lightweight portable chargers are Suntactics' sChargers which are small solar panel devices used for emergency disasters situation, power outages and in remote areas. These chargers can be used to solar charge 5-V devices including Android phones, all Apple iPhone models, all Samsung Galaxy smartphone models, all iPods, Blackberry, HTC One, Google Nexus, Nokia Lumia, Nooks and other eReaders. Suntactics' sCharges are also used to power other electronic devices such as Canon, Nikon, GoPro cameras, GPS devices, Bluetooth speakers, USB batteries (Anker), AA/AAA battery chargers, and anything else that can be charged from an USB port.

Suntactics' sChargers come in three different sized models. The sCharger-5 is a single port 5-V 5-W panel with output of 1Amp/5V. It measures 6 × 6 inches closed, 6 × 11 inches open and weighs approximately 8oz, as shown in Fig. 1.3A. The sCharger-8 is a dual port panel with output of 1.6Amps/5V. It measures 7.75 × 7.25 inches closed, 7.75 × 14.5 inches open and weighs 14oz. The sCharger-14 is a dual port 5-V, 14-W panel with output of 2.8Amps/5V. It measures 11.5 × 7 × 1/4 inches closed, 11.5 × 14.5 × 1/8 inches open and weighs 1.4lbs. All of these chargers are waterproof up to 40 feet and contain mounting grommets, brackets, and flexible tripods. The brackets and tripods helps the sChargers wrap around tents, backpacks, poles, chairs, tree branches, and also to point and aim them directly toward the sun for maximum efficiency. The Suntactics sChargers are equipped with a unique "Auto-Retry Technology," which enables them to immediately start charging within minutes after being temporarily blocked from the sun. The charging of mobile devices using Suntactics takes the same time as charging from a wall outlet. These chargers are rugged, capable of sustaining critical damage, and can withstand most extreme weather conditions.

All the three models mentioned above come with emergency kits for power outages or emergency disaster situations. Both the sCharger-5 and sCharger-8 emergency power outage kits as shown in Fig. 1.3B, include an ultra-bright 10 LED light with flexible wire lasting 50,000 hours, a 7800mAh USB battery, a carrying case and the sChargers. The sCharger-5 emergency power kit can be used in cloudy weather to trickle charge energy into 7800mAh battery which can then be used to power the LED for 24hours of

(A)   (B)

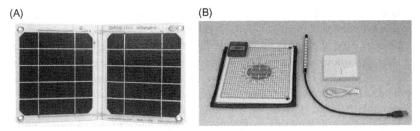

FIGURE 1.3   (A) Suntactics' sCharger-5 (B) sCharger-8 emergency kit.

light. The size of the sCharger-5 emergency kit is approximately the size of a CD case which can be folded and easily carried. The sCharger-8 emergency kit is capable of capturing 1.6 Amps of power from direct sunlight which can then be used to charge 7800 mAh battery and a device such as camera or smartphone simultaneously. The dual port sCharger-8 can charge four USB devices simultaneously, which is very beneficial for power outages in the aftermath of disasters and remote areas. The sCharger-8 emergency charger is the size of a mouse pad and can be folded for carrying purposes.

The sCharger-14 emergency power kit includes two ultra-bright 10-LED light with flexible metal wires, two 7800 mAh USB batteries, and a carry case. It is capable of charging six devices. The sCharger-14 USB solar charger is more powerful than the other two models discussed above especially in cloudy and dreary weather conditions. The sCharger-14 is capable of capturing more than 2.8 Amps of power from direct sunlight and in cloudy weather, it can still charge a phone rapidly. The charger can charge two devices simultaneously from the sun. The sCharger-14 can solar charge two devices at the same time through its dual ports or charge two USB batteries, used to power a smartphone, tablet and for powering LEDs for many hours. The sCharger-14 kit is a reliable power source for power outages in the aftermath of natural disasters as well as proving power to the people off the grid in remote areas

- *Features of solar battery chargers*
  - *Time taken to charge a battery.* The time it takes to charge a battery depends on the capacity of the battery to be charged and intensity of sunlight incident on the solar panel. It may also depend on the level of the battery discharge. Charging time can be calculated by the following equation:

$$\text{Time taken to charge the battery} = \frac{\text{Capacity of battery in mA or A to be charged}}{\text{Output of solar panel in mA or A}}$$

  - *Weight.* Weight depends on the size of the solar panel. Increase in solar panel output requires larger size and increased weight. For portable battery charger around 1 pound of weight is ideal.

- *Ease of Use.* To obtain best results from the battery charger, it is important to meet optimum sun conditions. One way of achieving this is to use self-standing solar panels that can be adjusted to collect full sun. For ease of use, a universal outlet like a Universal Serial Bus (USB) hub will help to eliminate the need of a nonstandard connection.
- *Compatibility.* For compatible battery chargers, it is important to include adaptors which fit different connectors used in the portable devices such as cell phones, laptops, cameras, and certain medical devices. The most common connectors include USB, mini USB, and iPhone. It is also desirable to have multiple outlets to charge more than just a single device.

1.6   ENERGY NEEDS IN REMOTE AND OFF-GRID AREAS

A community is said to be designated as "remote" if it is not connected to a central energy infrastructure such as a national electricity grid or a natural gas pipeline. It is the lack of connection to central distribution *infrastructure* which determines the remoteness of a community and not the lack of connection to road networks or seaport. A community without a connection to central energy infrastructure has to depend on liquid fuels such as diesel or petrol that may be delivered by land, air, or sea in the absence of local energy resources. Liquid fuel can be costly, environmentally unsafe, noisy, maintenance-intensive and creates dependency on imported energy. In recent years, other cleaner forms of power generation such as solar has become practically feasible, although other forms like wind and water power are also becoming more common. These forms of energy are renewable, unlike diesel or petrol, and are clean, maintenance free, and cost effective for remote communities with no access to grid. Examples of off-grid areas where electricity is not available are:

- Disaster and emergency sites
- Remote areas in developing countries
- Farms and ranches in the developing countries
- Homes in remote locations
- Businesses in remote locations
- Industrial plants in remote areas including mining
- Oil exploration including oil pump jacks
- Telecommunications including cell towers
- Water wells and irrigation
- Camp (cabins) and picnic sites

1.7   ENERGY NEED OF REMOTE AREAS IN DEVELOPED COUNTRIES

Off-grid photovoltaic was first used for practical applications in the industrialized countries of North America and Europe, where it was invented and developed. Part of the reason for its use in United States was the attraction that the customers with a highly individualistic culture have to generate electricity for personal use. Currently, PV has become a practical technology for remote areas of the developed world where the power needs previously were met, if at all, through diesel generators. Off-grid power systems can be more cost effective than connecting to the grid in remote locations where extending the grid can cost between US$15,000 to US$50,000 per mile depending on the climate, bedrock, and other environmental conditions. In some developed countries with geographies that limit grid penetration to remote areas, the use of off-grid solar is increasing. In many areas of United States and developed countries where damaging weather events are becoming more frequent, people are adopting grid tied or off-grid solar systems to supplement their utility electricity and provide backup power during the outages.

American farmers were one of the first users of off-grid photovoltaic power systems in remote locations around the farms and ranches. The reason for using off-grid PV systems for a number of low power agricultural needs when running transmission lines from the utility was either too expensive or not possible. In the absence of grid, the traditional usage of diesel, kerosene, and propane in agricultural operations presented problems such as noisy generators, cost of transporting fuel and fuel spillage, high maintenance needs, hazards and damage to environment. The use of photovoltaic power systems decreases pollution, has low maintenance costs, is simple to maintain, and offers cost effective system to produce on-farm energy for agriculture applications. The applications include water pumping for irrigation, watering livestock, pond management as well as battery charging, lighting, electric fencing, and the control of sprinklers, sprayers, and feeders around the farm. Most of these applications are considered to be remote and maintenance free for which photovoltaic is an excellent alternative. A survey by USDA in 2011 for solar energy production in 2009 indicated that solar panels are the most prominent way to produce on-farm renewable energy. It also indicated that production of solar energy occurs in every state of the union with 93% of farms equipped with on-farm renewable energy production.

In the United States, various national organizations since the landfall of Hurricane Sandy on Oct. 2012 have been focusing on enhancing the resilience of their electrical infrastructure against extreme events by the use of renewable energy solutions. These concerns are gaining foremost attention in remote areas of United States where reliability and stability of electrical supply is essential because of its isolation. The increased accessibility of renewable energy such as photovoltaic, wind and hydro power has led to an interest in integrating them to remote electric grids. The location of these electric grids in United States include military bases and national parks as well as places like American Samoa, Guam, Hawaii, Northern Mariana Islands, Puerto Rico, US Virgin Island, mainland communities in Alaska and communities at the coast of Maine and Massachusetts.

In Canada, there are over 630 First Nation governments serving a population of over 700,000 people, with the majority having no access to the electric grid because of their remote location. These communities are currently using environmentally-damaging diesel generators for their electrical needs and need clean and safe energy. Diesel generators produce air pollution, contribute to unhealthy smog and are also expensive and unreliable. Some First Nation communities have initiated the use of renewable energy in the form of solar, wind and hydroelectric energy. The Government of Canada, in partnership with private industry, has started to collaborate with First Nation remote communities to provide access to clean and affordable energy.

In Australia, 30 remote off-grid communities in the Northern Territory (NT) will receive the benefits of solar energy to displace the use of costly diesel generation for electricity generation. The funding was announced in 2014 by the Australian Renewable Energy Agency and NT government which will be managed by Power and Water Corporation through its subsidiary Indigenous Essential Services. The Northern Territory is a federal Australian territory in the center and central northern regions and is inhabited by 243,700 people of more than 100 nationalities. Many of the indigenous people, a quarter of the population of the area, still live in remote areas with their traditional lifestyle. The purpose of the project is to construct 10 MW of solar power in 30 remote off-grid communities to displace diesel fuel for electricity generation. This will be achieved by installing a 9 MW power system to displace 15% diesel fuel for electricity generation in majority of remote communities and 50% diesel displacement in another site by using 1 MW PV system using advanced cloud forecasting and energy storage in another site. The project aims to show that by integrating photovoltaic energy with existing diesel power systems will provide the off-grid remote community with low cost electricity and help to reduce the hazards of diesel generation and its damaging effect on the environment.

1.8 ENERGY NEED OF REMOTE AREAS IN THE DEVELOPING COUNTRIES

According to the International Energy Agency (IEA) estimate, developing countries will need to double their electrical power output by 2020. In addition, many developing countries that use existing generating capacity do not have enough to meet the current demand. This has resulted in frequent blackouts in many cities especially due to continuous process of urbanization from rural areas. The electrical utilities find it difficult to cope with this increased demand and as a result, it will take much longer for grid to reach rural areas in many developing countries. This has necessitated finding alternative solutions to provide electricity for the remote rural areas. The first alternative is to use diesel generating sets to provide electricity for local areas. This alternative, although widely used, creates polluted air, is costly and can be hazardous if not used properly. The second alternative is to use renewable energy technologies in the form of decentralized power generation.

It is estimated that in 2011, 1.3 billion people or 18% of the world population did not have access to electricity. It is also estimated that 80% of these people live in South Asia or Sub-Saharan Africa and remaining 20% reside in central Asia, central and south America as well as in urban areas in developing nations as shown in Fig. 1.4.

According to the IES, two-thirds of the population in the urban areas of developing countries since 2000 have gained access to electricity, while the population without access to electricity has become more concentrated in the remote areas. According to the World Bank, the number of persons with no access to electricity is likely to increase in the next few decades unless more effective policies are initiated to speed up off-grid electrification and the expansion to existing utility grids in to remote areas. The developing nations, according to IEA estimate, over the next 20 years will experience an average annual growth rate of 4% in primary energy use while the global demand for energy will grow approximately 60% by the year 2030. It is also expected that energy demand in Asia will double over the next 25 years driven primarily by population growth and economic power. Currently, the number of people without access to energy services in the world's rural population is estimated to 56% and the number of people without access to modern energy services will increase by 30 million every year.

In the absence of access to a grid, many people in the developing countries obtain their household energy for cooking and heating from solid fuels consisting of wood, charcoal, coal, animal dung and crop waste burned either in

	Without access to electricity		Traditional use of biomass for cooking*	
	Population	Share of population	Population	Share of population
Developing countries	1257	23%	2642	49%
Africa	600	57%	696	67%
Sub-Saharan Africa	599	68%	695	79%
Nigeria	84	52%	122	75%
South Africa	8	15%	6	13%
North Africa	1	1%	1	1%
Developing Asia	615	17%	1869	51%
India**	306	25%	818	66%
Pakistan	55	31%	112	63%
Indonesia	66	27%	103	42%
China	3	0%	446	33%
Latin America	24	5%	68	15%
Brazil	1	1%	12	6%
Middle East	19	9%	9	4%
World***	1258	18%	2642	38%

FIGURE 1.4 Number of people without access to modern energy services in million by region, 2011. www.worldenergyoutlook.org/resources/energydevelopment. *Based on the World Health Organization (WHO) and IEA databases. **Since *WEO-2012*, population numbers for India have undergone a significant upward revision, meaning that the electrification and clean cooking access rates have not changed, the number of people estimated to be without access has significantly increased. ***Includes OECD countries and Eastern Europe/Eurasia.

open or in traditional stoves. Burning of such a solid fuel using an inefficient traditional stove leads to incomplete combustion and high levels of indoor and outdoor air pollution emission and poor air quality. The emission from burning of solid fuel in a traditional stove has a significant effect on global warming because of incomplete combustion of carbon which emits pollutants such as mercury compounds, nitrogen oxide and sulfur dioxide in addition to arsenic, mercury, fluorine, and sulfur which makes solid fuel dangerous to cook with. Cooking with solid fuels also produces greenhouse emissions with direct CO_2. This accounted for nearly one-fifth of global CO_2 emission in 2008. The net contribution of incomplete combustion of solid fuels including greenhouse gases (methane) and short-lived pollutants (black carbon) is much greater and have the most immediate effect on climate change. Methane emissions are considered to be second largest cause of climate change after carbon dioxide. It is also dangerous for human health.

According to World Bank, nearly two million people die every year from chronic lung disease, lung cancer and pneumonia due to exposure to indoor pollution resulting from cooking with coal, wood and biomass. Almost 99% of these deaths occur in developing countries with 44% of these deaths in children and 60% deaths in women who are also vulnerable to attack and injury during fuel collection. Recent study has also shown increased risk of this population with adverse pregnancies, cataracts, cardiovascular diseases, and tuberculosis.

The safe access of electricity to people in remote communities could lead to life-saving improvements in health, education agricultural operations, communications, and clean water. It will also help to build a sustainable local economy by creating entrepreneurial opportunities. Almost 85% of people without access to electricity live in the remote rural areas or on the fringes of metropolitan cities. Extending energy grids to these remote areas is expensive and time consuming. According to a United Nation estimate, investments of US$35-40 billion a year is needed until 2030 for each person in the world to cook, heat and light their living places and have energy for other productive uses. However the IEA estimates that based on current trends, 16% of the world's population will still have no electricity by 2030. Thus instead of using a top-down approach and relying on governments and utilities, a local bottom-up approach is needed for a sustainable and carbon free systems in the remote areas. This can be achieved by making use of photovoltaic technology which has become more efficient and less costly in the recent years. In the developed countries photovoltaic energy is becoming a source of supplementary grid power and also helping to enhance the resilience of the grid by moving toward flexible distributed system. Developing countries have an opportunity to use it in remote areas applications of domestic lighting, battery charging, water pumping and medical procedures in rural clinics as well as leapfrog the outages-prone centralized system. The use of photovoltaic energy does not emit any greenhouse gases or smoke. The sun as a fuel is available free of charge which makes it cost effective for communities with no access to grid in the remote areas of the world. Use of direct solar thermal energy can also be used to power cooking stoves which do not produce smoke, hence eliminating health related hazards associated with cooking over open fires or traditional stoves.

1.9 PHOTOVOLTAICS FOR DISASTERS RELIEF AND REMOTE AREAS

Although there is no electrical source that is completely immune from natural disasters, the use of photovoltaic and fuel cells, if properly designed, can be reliable and safe. These distributed energy resources are cleaner, smaller, and produce power directly at the source that can operate independently when the grid malfunctions due to disaster or shortage of fuel. These technologies, although not completely safe, can provide more reliable power to run critical loads at full capacity. In the case of providing backup power in the aftermath of a disaster, the biggest challenge is to provide enough power for longer periods whether using a battery system or combustion-based generators. Photovoltaics with battery storage is affordable and one of the most cleaner and reliable source of power for longer period of time for homes. Electricity from PV can increase the resilience and help to respond and recover from disasters by powering medical devices, communications, water purification and water pumping, and refrigeration equipment as well as lighting the premises. The use of PV power helps to replace portable diesel and gasoline generators which are expensive, hazardous, damaging to the environment and rely on scarce fuel in the aftermath of disasters. In addition, these systems can also be used to power critical infrastructure in remote areas that have no access to electric grid.

Photovoltaic energy has recently been used for disaster relief by National Renewable Energy Laboratory (NREL) and Florida Solar Energy Center (FSEC) who developed PV systems and used to power equipment for emergency in the aftermath of disasters. NREL also provided technical advice and guidance to Federal Emergency Management Administration (FEMA) on how to educate their staff on the applications of photovoltaics for disaster relief and for building disaster-resistant communities. FSEC provided the first portable PV systems for disaster relief after Hurricane Hugo (1989), introduced to emergency management organization by members of PV industry. This PV system was used to power 12 V DC fluorescent lights, Ham radios and fans at various disaster shelters, medical facilities and emergency management offices. The trailer-mounted systems have also been used in many disasters including Northridge earthquake (1994), Hurricane Andrew (1992), Hurricane Bonnie (Aug. 1988), Hurricane Georges (Sep. 1988), Hurricane Charlie (2004) and Hurricane Katrina (2005). In 2010, 10 solar trailers were developed and used after the Haiti earthquake by a team of IEEE volunteers.

In 2010, the FSEC implemented the SunSmart Emergency Shelter program through a contract awarded by the DOE-administered *American Recovery and Reinvestment Act* to install PV systems at the shelter-designated schools for power outages. Initially the program started as a way to raise awareness and understanding of photovoltaic technology among students, teachers and the general public. The SunSmart E-Shelter program, coordinated by the FESC has installed more than 115 PV array systems of 10 kW capacities each at the emergency shelter designated schools in Florida, United States. Each PV system consist of 10 kW ground mounted solar panels (SolarWorld) connected to inverter/charger (Outback's Flexware 500) and 25 kWh storage battery (Sun-Xtender) for backup power. The system has bimodal configuration and is used to power critical loads identified by emergency organization such as phone chargers, radios, overhead lights, and life-support equipment during and in the aftermath of a disaster. It is also integrated into the shelter part of the school. The participants of the program can also make use of a disaster kit consisting of radio, phone and battery charger, flashlight and DC inverter which can be bought from a camping store. These systems are connected to the grid, hence any excess energy is usually sold to utility companies. In recent years, these E-shelters have become more disaster-resistant and comfortable for the people displaced by the disasters after the revised FEMA standards. The SunSmart E-Shelter Program has added more than 1 MW of combined photovoltaic generating capacity to Florida using American-made components. In addition, through the education and outreach efforts of SunSmart E-Shelter program, more than 350 Florida teachers have received professional development in the use of photovoltaic and more than 50,000 students have been introduced to photovoltaic and renewable energy. One of the primary objectives of SunSmart E-shelter program, to make an "active solar community," through teacher and student education has become a reality in 115 school across Florida. A companion inquiry based program rich in STEM content has also been designed to explore; (1) How photovoltaic work; (2) What impact the system has on their school; (3) Other alternative energy sources; (4) How to prepare for disasters; (5) Relationship between energy and the environment, history, geography, economics and art. The success of SunSmart program was made possible by the strong partnership of US Department of Energy, the University of Central Florida/Florida Solar Energy Center, Florida solar industry, utilities and the state of Florida who funded, advised, and supported in executing the program at different phases.

In the aftermath of 2012 Superstorm Sandy, the volunteer-led "solar Sandy Project" stepped in with a mobile solar generator to help the hardest-hit communities along the New Jersey shore and in Rockway Beach, Queens, and Staten Island. Solar generators of 10 kW were used to warm food, power mobile phones and laptops, and recharge power tools and other critical equipment.

In Thailand, a large PV power system survived with no damage to the solar panels after the great floods of 2011. Sunny Banchak, a 45 MW project, consisting of 157,200 PV modules located at 60 miles from Bangkok remained under fresh water for nearly 2 month. The panel suffered some solar field damages but produced 8 MW of output after 6 months and resumed full production after 9 months. The fully operational Suntech solar panels were replaced after the flood and were purchased by World Machine Center Co., Ltd. who rehabilitated the panels and sold to farmers, temples, families, and corporations in Thailand to get access to inexpensive and reliable electricity.

A novel method of providing electricity to remote areas and disaster relief has been developed by SunDial which was first deployed in Afghanistan during a 2010–12 Special Forces operation. The system uses a 20-foot shipping container consisting of 120 installation-ready solar panels, a combined communication center and accessories to be transported to remote places. This mobile unit produces 28.8 kW at peak daylight hours and can be increased to 34.2 kW power. During sunlight, solar panels power the load and also charge the 64 storage batteries housed in the trailer. The batteries provide power at night and an optional backup diesel generator kicks in automatically when the batteries become depleted. During the day the diesel generator turns off and the solar panels take over which produce power and recharge the batteries. The mobile PV power system is currently used for powering a water purification facility in the remotest part of Nigeria to replace a diesel generator at a reduced cost. It is also used in Asia, Africa and North America for providing electricity to remote locations for disaster relief, rural electrification, poor grid access or military purposes. The empty container can be used as an internet cafe, telemedicine clinic, a corporate office, or a place to live. The mobile system was initially developed for the US military zone in the remote areas where transporting diesel fuel to run their operation is dangerous, costly, and hazardous.

1.10 PHOTOVOLTAICS AROUND THE WORLD

It is predicted that photovoltaic market will continue to expand around the world and increase its share of the energy mix because it provides clean, safe, decentralized, and affordable energy to people in the aftermath of disaster and remote areas. This is evident by the record increase of 38.4 GW in 2014 of newly installed PV systems and a 138.9 GW of cumulative capacity worldwide in 2013 despite its recent economic crisis. World solar power in 2014 is equivalent to installing 12 nuclear power plants. This is worth considering that America's largest nuclear plant near Phoenix, Arizona, has a capacity of 3.7 GW which provides power for 4 million people in Southwest and Southern California.

This increase in newly installed PV systems is attributed to China which was the top market with 11.8 GW, followed by Japan with 6.9 GW and the United States with 4.8 GW. According to the US Energy Information Administration's Apr. 2014 update, numbers of installed solar photovoltaic in United States has grown by 418% between 2010 and 2014. In Europe, new PV installation capacity amounted to almost 11 GW with Germany 3.3 GW as a top market, resulting into a significant decrease from the previous years. However, the number of new PV installations in 2013 was the highest in Europe after the use of windmills as the electricity source. There were more sources of energy such as gas which experienced net negative number with more capacity decommissioned than installed. The size of European PV market in 2013 remained stable with around 6 GW per year in the last 3 years despite its decline in Germany and Italy. In 2009, a study noted that most of the PV development took place during the period when state subsidies were made available to countries such as Germany, South Korea, Spain and the United States. However the recent declining political support to these subsidies has created a climate of uncertainty that will obstruct the redevelopment of the PV markets. But outside Europe, the potential for PV market continue to grow and could transform into real market take-off in many countries including India with 1115 MW, Canada with 444 MW, South Korea with 442 MW, and Thailand with 317 MW. The implementation of new feed-in-tariff policies in countries such as China and Japan has led to significant increase of PV markets. As in most countries, PV remains a policy-driven market; declining political support for PV has led to reduced markets in several European countries such as Belgium, France, Germany, Italy, Spain and others.

The phenomenal growth in PV markets over the past years was driven by cost reduction of PV technology as well as state support in the form of subsidies, incentives, tax credits and feed-in-tariffs predominately in developed countries with relatively modest solar irradiation. However, due to declining political support in most European countries, the market growth is shifting toward China and South Asia, followed by South America, Middle East and North African countries. According to EPIA North America, China and India will drive the market after 2020 while Africa, the Middle East and South America will also provide significant contributions after 2030. In the Sunbelt countries located within 35 degrees around the equator with exceptional solar irradiation, PV potential remains untapped. EPIA found that 66 countries out of 148 countries that compose the Sunbelt countries currently represent only 9% of global installed PV capacity. It is estimated that PV potential of Sunbelt countries where PV can compete with diesel generators for peak power generation without financial support could range from 60 to 250 GW by 2020 and from 260 GW to 1.100 TW in 2030. This will amount to 2.5–6% of the Sunbelt's overall power generation which will bring electricity to around 300 million people. According to an EPIA report, Sunbelt countries currently represent around 75% of the world's population and 40% of global electricity demand. With the exception of China, all the top 10 major PV markets in the world are currently located outside the Sunbelt in modest solar irradiation. It is also predicted that in the next 20 years, 80% of forecasted growth of the world electricity demand will originate from the emerging economies of Sunbelt countries.

The developed countries will be dominated by the grid connected system while the developing countries will have the combination of both grid as well as off-grid connections. The grid connected systems in developing countries will be integrated into the cities and town's electricity networks while off-grid and minigrid systems will be used to power villages and remote areas. There are 500,000 to 1,000,000 off-grid PV systems which are being used in developing countries. The commonly used PV system is a small solar home system of 30–100 peak power. This system consists of a solar panel, a storage and accessories which is used to power lights, radio, fans, and cell phone chargers. A bigger system is also used for powering street lights, and water purifiers both for AC and DC applications.

The off-grid installation of solar systems normally used in remote areas are also taken into account for the total installation in countries such as Australia and South Korea where dozens of megawatt of power is installed every year. Other countries like United States also include their off-grid systems which were 10% of overall market in 2009 but has declined since. In countries such as Peru and India, the development of PV in the coming years could originate partially from their newly started hybrid system and microgrid applications.

1.11   GROWTH AND FORECAST OF PHOTOVOLTAIC MARKETS

The growth of PV markets creates new jobs which in turn benefits the country's economy. According to International Renewable Agency (IRENA), the number of people working in the global renewable industry grew by 14% to 6.5 million people in 2013. The solar panel industry employed the largest number of people, 2.27 million worldwide followed by biofuels, 1.45 million, and wind power 830,000. The largest employers of renewable energy are China with 2.6 million people, followed by Brazil with 890,000, and the United States with 630,000 people. Two-thirds of the 2.6 million people employed in China are in the solar industry which has increased three time in 2 years. Brazil and United States also have large sector of people employed in biofuel. The main reason behind the rise in jobs is the increased installation as a result of decreased cost of PV systems in countries such as China, Japan, United States and other.

According to The Solar Foundation Fifth Annual National Solar Jobs Census, the US Solar industry employed nearly 174,000 people in 2014 with a 21.8% growth since Nov. 2013. The census also showed that solar industry employment grew nearly 20 times faster than national average employment growth rate of 1.1% in the same period. The people working in solar industry are employed at nearly 6100 businesses in different US states. In dollar amounts the PV businesses increase every year and solar installation valued at US$13.7 billion in 2013, against US$11.5 billion in 2012 and US$8.6 billion in 2011. In the last 4 years the solar industry added nearly 81,000 jobs with a growth of 86% in the solar sector. In United States over the last year nearly 1.3% of all jobs were created by the solar industry and it is expected that it will add 36,000 jobs in 2015. Most of these jobs are in the solar installation sector which is larger than fossil fuel generation such as coal mining. In the solar installation sector, 50% more jobs were added than the total jobs created by both the oil and gas pipeline construction industry and the crude petroleum and natural gas industry in 2014. This shows that growth in the solar industry is putting people back to work and the solar industry has proven to be a strong engine of economic growth and job creation.

According to recent report, India's solar market has grown more than hundred times in the last 4 years and it is the fifth-largest wind energy producer country in the world. The combined solar and wind energy industry jobs are estimated to be more than 70,000 which are considered as clean energy jobs. Approximately 24,000 jobs were created by PV projects commissioned between 2011 and 2014. The majority of these jobs were created for the construction and commissioning of larger PV plants, although smaller projects of building up to 5 MW may provide higher number of jobs per MW. It is expected that more than 220,000 jobs will be created in commercial rooftops and 320,000 jobs in the residential rooftops in the next decade. The number of jobs in utility-scale solar and ultra-mega solar plants will be less than 134,000 during the same period. Many US companies and NGOs are currently working to help and receive the benefits of India's expanding market opportunities for clean energy. Some organizations, including USAID has come up with innovative financing solutions such as microfinancing and crowdfunding to help Indian NGOs and other to realize the full potential of renewable energy. Leaders of countries with renewable energy feel that growth of clean energy is creating new jobs for their people, enhancing access to energy and reducing the country's climate impact on the world. The United States is collaborating to ensure India stays on their path for clean energy.

The PV market in many countries remains a policy-driven market where political decisions influence potential market growth or decline. One example of this is Germany which dominated the worldwide solar market due to extensive government subsidies for many years. However, Germany as well as Italy is gradually reducing their feed-in-tariffs affecting the decline of annual newly installed markets. This, however, is not a reduction in the world newly-installed solar markets because the rest of the European countries, Africa, Middle East and the Sunbelt countries will take up this slump in growth. China and Japan still have feed-in-tariff programs and together with the United States are seeing dramatic growth. Japan and Germany intends to phase out their nuclear generators and replace them with solar systems because of recent accidents in nuclear power stations. To meet the increasing demands, India plans to install less-costly solar plants instead of building fossil fuel plants with transmission lines. There is also considerable opportunity for growth in other regions, led by Chile, South Africa, and Saudi Arabia. Many countries are trying to achieve solar generated electricity prices close to grid parity.

Many analysts believe that after 2014, the long-term growth for PV markets are expected to normalize and will be between 20% and 25%. Navigant Research, in late 2013, predicted that annual revenue from PV installations will surpass US$134 billion by 2020. Frost and Sullivan also predicted in 2014 that the PV market will increase to US$137 billion. It is further predicted that in 2020, PV will be cheaper and more popular because it is also becoming a commodity. It is expected that in 2020, photovoltaics will be cost competitive with retail electricity prices without subsidies, incentive or tax credits, in many regions of the world.

BIBLIOGRAPHY

[1] N.C. Abi-Samra, One Year Later: Superstorm Sandy Underscores Need for a Resilient Grid, IEEE Spectrum. <http://spectrum.ieee.org/energy/the-smarter-grid/one-year-later-superstorm-sandy-underscores-need-for-a-resilient-grid>, 2013.

[2] B. Ball, "Rebuilding Electrical Infrastructure along the Gulf Coast" A Case Study: A Proactive Approach to Disaster Preparation Is Crucial to Disaster Recovery. National Academy of Engineering of the National Academies. <https://www.nae.edu/Publications/Bridge/TheAftermathofKatrina/RebuildingElectricalInfrastructurealongtheGulfCoastACaseStudy.aspx>.

[3] D. Carrington, Solar Industry Leads 14% Rise in Renewables Jobs. <http://www.theguardian.com/environment/2014/may/12/solar-industry-renewables-jobs-china-us>, 2014.

[4] S. Dahlke, Solar Home Systems for Rural Electrification in Developing Countries. An Industry Analysis and Social Venture Plan. <http://www.csbsju.edu/Documents/CSB%20Sustainability/Solar%20Paper.pdf>, 2012.

[5] The Weather Channel, Cost of Natural Disasters Has Quadrupled in Recent Decades, Official Says. <http://www.weather.com/science/environment/news/cost-natural-disasters-has-quadrupled-recent-decades-official-20140606>, 2014.

[6] W. Flanagan, Battery Storage with Solar PV: The Next Logical Step. Renewable Energy. <www.renewableenergyworld.com/rea/news/article/2014/09/battery-storage-with-solar-pv-the-next-logical-step>, 2014.

[7] FEMA (Federal Emergency Management Agency), Listing of Emergency Management Higher Education Programs. <http://training.fema.gov/emiweb/edu/collegelist/EMMasterLevel/>, September 2013.

[8] Frost & Sullivan: Global Solar Market Continues to Rise and Shine, <http://www.frost.com/prod/servlet/press-release.pag?docid=291574892>, 2014.

[9] H.J. Gibbons, Physical Vulnerability of Electric Systems to Natural Disasters and Sabotage. U.S. Congress, Office of Technology Assessment, Physical Vulnerability of Electric System to Natural Disasters and Sabotage, OTA-E-453. <http://ota.fas.org/reports/9034.pdf>, 1990.

[10] IEA-Retd Renewable Energies for Remote Areas and Islands (Remote), Trama TecnoAmbiental- Meister Consultants Group—E3 Analytics HOMER Energy. <http://iea-retd.org/wp-content/uploads/2012/06/IEA-RETDREMOTE.pdf>, 2012.

[11] B. Kahn, Weather Disasters Have Cost the Globe $2.4 Trillion. Published: <http://www.climatecentral.org/news/weather-climate-disasters-cost-trillions-17773>, 2014.

[12] C. Keeves, New Report: India's Expanding Solar and Wind Energy Markets Jumpstarting Job Growth: U.S.–India Collaboration on Financing and Innovation Key to Continued Success. Environmental News: Media Center. <nrdcinfo@nrdc.org>, 2014.

[13] K. Kowalenko, Lighting Up Haiti, IEEE Volunteer Help Bring Electricity to Rural Areas. The IEEE Institute. <http://theinstitute.ieee.org/benefits/humanitarian-efforts/lighting-up-haiti>, 2011.

[14] J. Kever, NRG Vehicle Uses Solar Power to Provide Disaster Relief. <http://www.houstonchronicle.com/business/energy/article/NRG-Vehicle-uses-solar-power-to-provide-disaster-4608671.php>, 2013.

[15] E. Luo, Solar Energy Can Mitigate Impact of Weather-Related Disasters. <http://www.suntech-power.com/news/news130.html>, 2014.

[16] J. Leaning, D. Guha-Sapir, Natural disasters, armed conflict, and public health, Natl. Eng. J. Med 369 (2013) 1836–1842 http://www.nejm.org/doi/full/10.1056/NEJMra1109877#t=article.

[17] J.J. Lee, C. Layrent, C. Becker-Birck, Solar PV emergency & Resilience Planning, Meister Consultant Group. <http://solaroutreach.org/wp-content/uploads/2013/08/Solar-PV-Emergency-Resilience-Planning_Final.pdf>, 2013.

[18] T. Murphy, Do the Math and Estimation to Assess Energy, Growth, Options Using Physics. <http://physics.ucsd.edu/do-the-math/2011/11/a-solar-poweredcar/#sthash.9nJHTWPg.dpuf>, 2011.

[19] L. Mearian, Ford Builds Solar-Powered Car. Computer World. <http://www.computerworld.com/s/article/9245116/Ford_builds_solar_powered_car>, 2014.

[20] Navigant Research, Solar PV Market Forecasts Installed Capacity, System Prices, and Revenue for Distributed and Non-Distributed Solar. <http://www.navigantresearch.com/research/solar-pv-market-forecasts>, 2014.

[21] B. Oral, F. Dönmez, The Impacts of Natural Disasters on Power Systems: Anatomy of the Marmara Earthquake Blackout. <http://www.uni-obuda.hu/journal/Oral_Donmez_23.pdf>, 2010.

[22] S. Prabhu, India's Burgeoning Solar and Wind Energy Markets Jumpstarting Job Growth. India's Burgeoning Solar and Wind Energy Markets to Boost Employment. <http://www.nrdc.org>, 2014.

[23] D.S. Rice, Report: Climate change behind rise in weather disasters. USA Today, (2012).

[24] Santactics Portable Solar Chargers. <http://www.suntactics.com>.

[25] O. Schäfer, Global Market Outlook for Photovoltaic 2014–2018. European Photovoltaic Industry Association. <http://www.epia.org/fileadmin/user_upload/Publications/EPIA_Global_Market_Outlook_fPhotovoltaics_2014-2018_-_Medium_Res.pdf>, 2014.

[26] Z. Shahan, The Solar Market Continues to Defy Forecasts. <http://www.abb-conversations.com/2014/09/the-solar-market-continues-to-defy-forecasts/>, 2014.

[27] SolarBuzz, Solar PV Industry Targets 100 GW Annual Deployment in 2018. <http://www.solarbuzz.com/news/recent-findings/solar-pv-industry-targets-100-gw-annual-deployment-2018-according-npd-solarbuzz>, 2014.

[28] Solar Markets. <http://solarcellcentral.com/markets_page.html#>.

[29] Solar Generation 6, Executive Summary. <http://www.greenpeace.org/seasia/ph/Global/international/publications/climate/2010/SolarGeneration2010.pdf>, 2010.

[30] Steady Increase in Climate Related Natural Disasters, <http://www.accuweather.com/en/weather-blogs/climatechange/steady-increase-in-climate-rel/19974069>, 2013.

[31] B. Stuart, Report: PV poised to become mainstream source of energy in Sunbelt countries. Research & Development, Market & Trends, Top News. <http://www.pv-magazine.com/news/details/beitrag/report--pv-poised-to-become-mainstream-source-of-energy-in-sunbelt-countries-_100001467/#ixzz3PbnQtzFS>, 2010.

[32] European Photovoltaic Industry Association (EPIA), Unlocking the Sunbelt potential of Photovoltaic. <http://www.epia.org/uploads/tx_epiapublications/Sunbelt_Epia_MARCH2011_FINAL.pdf>, 2011.

[33] A. Upadhyay, Beehives, Elephants, 220,000 Jobs, & the Future of Solar Rooftops in India. <http://cleantechnica.com/2014/11/16/beehives-elephants-220000-jobs-future-solar-roof-tops-india/>, 2014.

[34] U.S. Solar Market Insight, SEIA/GTM Research Solar Industry Data. <http://www.seia.org/research-resources/solar-industry-data>, 2014.

[35] S. Vorrath, Solar PV Market May Increase To 500 GW By 2018. Originally Published on RenewEconomy. <http://reneweconomy.com.au/>, 2014.

[36] M. Walker, Homer helps bring electricity to remote areas of Afghanistan, Microgrid News. <http://microgridnews.com/homer-helps-bring-electricity-to-remote-areas-of-afghanistan/>, 2012.

[37] White House Council of Economic Advisers and Energy Department Release New Report on Resiliency of Electric Grid During Natural Disasters. <http://energy.gov/articles/white-house-council-economic-advisers-and-energy-department-release-new-report-resiliency>, 2013.

[38] World Energy Outlook, International Energy Agency (IEA). <http://www.worldenergyoutlook.org/pressmedia/recentpresentations/London November12.pdf>, 2013.

[39] World Bank, Disaster Risk Management Overview. <http://www.worldbank.org/en/topic/disasterriskmanagement/overview>, 2014.

[40] World Bank, Managing Disaster Risk for Resilient Development. <http://www.worldbank.org/content/dam/Worldbank/document/SDN/Full_Report_Building_Resilience_Integrating_Climate_Disaster_Risk_Development.pdf>, 2014.

[41] M.I. Xiarchos, B. Vick, Solar Energy Use in U.S. Agricultural Overview and Policy Issues, United State Department of Agriculture, Office of Energy Policy & New Issues. <http://www.usda.gov/oce/reports/energy/Web_SolarEnergy_combined.pdf>, 2011.

[42] W.R. Young Jr., Photovoltaic Applications for Disaster Relief. FSEC-849-95, Florida Solar Energy Center, Cocoa, FL, 1995 and 2006.

[43] W.R. Young Jr., Developing Mobile PV Emergency Power System in a Disaster. ASES, Solar 2009, Buffalo, NY, 2009.

[44] B. Young, Renewable Energy to the Rescue. <http://solartoday.org/2013/03/renewable-energy-to-the-rescu>, 2013.

[45] 8 Innovative Emergency Shelters for When Disaster Strikes. Inhabitat. IKEA and UNCHR <http://inhabitat.com/8-innovative-emergency-shelter-designs-forwhen_disaster-hits/ikea-refugee-shelter2-2>.

Chapter 2

Fundamentals of Standalone Photovoltaic Systems

2.1 INTRODUCTION

Standalone PV systems are autonomous systems which operates independently from a utility grid. These systems provides affordable and reliable sources of electricity, and are best suited for remote areas where the grid is either nonexistent, unreliable, or is damaged due to a disaster. Standalone PV systems are also known as "off-grid systems" as they are not connected to the grid. In United States, there were nearly 180,000 families living off the grid in 2006 and this number is expected to increase to more than 300,000 in the coming years. There are also more than 1.2 billion (estimated) people with no access to electricity, most of them living in the remote areas of the developing world. It would not be economical to connect remote areas to the grid, considering the high cost involved and low power requirements of the people living in these areas. It is logical to use standalone PV systems which would be more cost effective under these circumstances. Standalone PV systems could also be used by individuals who wish to obtain independence from power-providing utilities or demonstrate a commitment to a pollution-free environment.

A standalone PV system supplies electricity to a consumer directly or through rechargeable batteries. Standalone PV systems can be assembled for temporary or permanent usage in a short time, and can replace gasoline and diesel generators in many applications. These systems were the predominant PV systems before the grid-tied systems became more popular in recent years. One of the biggest advantages of standalone PV systems is to use them in applications with outputs ranging from microwatt to megawatts such as watches, calculators, fans, water pumping, remote communications, satellites, space vehicles, and megawatt scale power plants.

2.2 TYPES OF STANDALONE PV SYSTEMS

Standalone PV systems can be classified into two categories: direct PV systems and battery storage PV systems.

Standalone Photovoltaic (PV) Systems for Disaster Relief and Remote Areas.
DOI: http://dx.doi.org/10.1016/B978-0-12-803022-6.00002-2

- A direct coupled standalone PV system consists of PV panels which produce electricity to power a DC load as shown in Fig. 2.1A.

  A direct coupled system is not connected to any storage batteries or grid and contains one or more solar panels, disconnect float switch, overcurrent protection and electrical load. These systems are used only during the daylight hours. Examples of a DC load include all electronic devices that can be powered by direct current such as LEDs, solar powered water pumps, fans, refrigerators, florescent lights, radios, and TVs. The critical part of designing a well-performing direct coupled system is to match the impedance of the electrical load to the maximum power output of the PV array. For certain loads, such as positive-displacement water pumps, a type of electronic DC–DC converter called a maximum power point tracker or a charge controller, is used between the array and the load to help better utilize the available array maximum power output. This is shown in Fig. 2.1B.
- Battery storage standalone PV systems

One of the disadvantages of direct coupled systems is their inability to operate during night time or on cloudy days with insufficient light. This disadvantage can be overcome by using a PV system with a battery storage. Such a system as shown in Fig. 2.2 will store power during the day when solar modules are producing electricity and use it during the night or cloudy days when solar modules are not producing electricity.

The use of a battery in battery storage systems is essential since the demand from the load does not always match the PV panel capacity. It provides the following tasks in the standalone PV system:

1. Stores electricity when excess power is available and provides it when needed
2. Provides surge currents to loads like electrical motors when needed
3. Provides stable current and voltage by eliminating transients

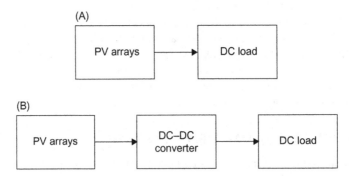

FIGURE 2.1 (A) Block diagram of direct PV system (B) Direct PV system with DC–DC converter.

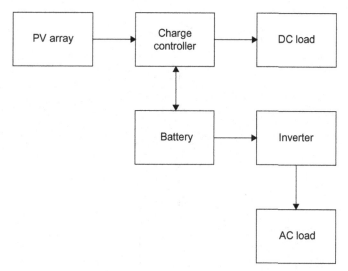

FIGURE 2.2   Block diagram of battery powered standalone system.

2.2.1   Battery Storage Hybrid Standalone PV Systems

In the case of some standalone PV systems when it is important that power is always available, the system may also have another source of power such as a diesel generator, a wind turbine, or biofuel. Such a PV system that uses a combination of two or more types of electricity production is known as a PV hybrid system and is shown in Fig. 2.3.

A popular combination of such a hybrid system is integration of a PV system with a diesel generator as a backup to reduce consumption of diesel fuel and to minimize atmospheric pollution. This hybrid system has also shown to be more economically viable than a standalone diesel generator system. As the generation capacity of diesel generator is limited to a certain range, it is a viable option to include battery storage to optimize the sun's fluctuating contribution to the overall generation of a hybrid system. Other PV hybrid systems include a combination of solar and wind turbine systems. The peak operating times for these electricity generation systems occurs at different times of the day and year which makes power generation of this hybrid system more constant with less fluctuation than each of the two component subsystems. The main disadvantages of these hybrid systems are extra cost and regular maintenance because of the requirement for extra equipment.

2.3   TYPES OF BATTERY STORAGE STANDALONE PV SYSTEMS

The size of PV array and battery in storage standalone PV system depends on individual requirement and application of the system. This system can be a portable such as a flashlight or cell phone charger that can be carried in a backpack,

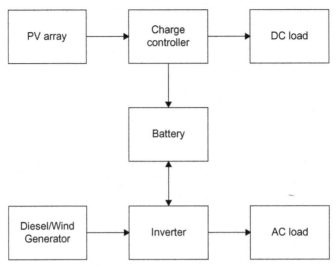

FIGURE 2.3   Battery storage hybrid standalone PV system.

or a solar suitcase for medical emergencies that can be carried by hand. It can also be much bigger in size, and needs to be moved by a trailer such as supplying power to a remote cellular base station or a water pumping system in remote areas. It can also be a self-reliant electricity grid capable of producing and distributing power within a small defined boundary for maintaining stable service. A microgrid can be powered by distributed power generators, batteries and/or renewable resources like PV arrays, wind turbines or biomass. Battery storage standalone systems can be divided into different categories based on their size, mobility and applications:

- Fixed standalone PV systems: These systems are essentially standalone PV systems but once installed, cannot be easily moved on wheels like mobile PV systems or carried manually like portable PV systems. Fixed standalone PV systems differ in size, capacity, and applications from mobile and portable PV systems. Examples of such systems include providing power for a remote villages, cellular base stations, water purification systems, remote medical clinics, irrigation and livestock watering and others.
- Mobile standalone PV systems: These systems usually have wheels to be able to be moved from one place to other. Mobile standalone PV systems include solar trailers which have been used to provide power in the aftermath of a disaster and remote areas. Other examples of mobile standalone systems include solar-powered clinics and shelters, solar-powered communication and water purification systems, solar-powered cars and solar-powered airplanes.
- Portable standalone PV systems: These systems are small in size and weight and can be carried manually by an individual. These systems include solar flashlights, cell phone chargers, solar briefcases, solar suitcases, or solar

backpacks. Their applications include providing light during power outages, camping or outdoor party in the remote areas or in the aftermath of a disaster. These portable systems have also been used for providing power to medical appliances, water purification systems and laptop computers in the remote areas.

- Standalone solar microgrids: Solar microgrids are autonomous electric grids that produce and distribute power across a limited area, such as a remote village, university campus, hospital, military base, industrial complex or an area disrupted by a disaster. Standalone microgrids are not tied to utility grid and there is no outside electrical connection. It is found that integrating PV systems with remote standalone microgrids will drive rapid growth of distributed power in rural and remote areas which are not accessible to traditional grid. This will also benefit small businesses, agriculture areas, communities in disaster-prone areas and the developing world with affordable, cleaner, and more reliable energy generation. According to Pikes research, developing countries will be the main beneficiaries where more than 80% of the world population live but consume only 30% of global energy production. In case of weather-related disasters, microgrids can strengthen grid resilience and help mitigate grid disturbances because microgrids are able to continue operating even when the main grid cannot.

2.4 ADVANTAGES AND DISADVANTAGES OF STANDALONE PV SYSTEMS

2.4.1 Advantages

- Independence from external energy supplies
- Savings for remote homes from high grid connection charges
- Minimal maintenance and low upkeep cost
- No byproduct and wastes
- Environmental friendly
- Easy expansion for additional PV panels and batteries
- Avoid escalating diesel generation costs

2.4.2 Disadvantages

- Reliance on sun which can be intermittent and unpredictable
- High initial investment particularly on PV panels and batteries
- Possible danger from battery acid fumes and spills

2.5 APPLICATIONS OF STANDALONE PV SYSTEMS

The following applications of standalone PV systems in remote areas may include fixed (stationary), mobile, and portable systems. These applications are also applicable to rapid deployment or temporary loads such as disaster, emergency, rescue, road repair and one-time events.

- Residential: Villages in remote areas, private homes and offices, power provision for houses and cabins, lighting, refrigerators, fans, TVs, radio, cell phone battery charging, automatic gate openers (significant applications in developing countries).
- Industry: Power tools, motors, battery charging, oil pipelines and other types of piping, provision of power for limited electric charges in the order of a few kW.
- Telecommunications: Mobile and cellular communication, microwave/radio repeaters, radio/television relay stations, telephone devices, stations for data surveying and transmission (meteorological, seismic, indicating the presence of fire and level of watercourses).
- Public Services: Drinking water and livestock water pumping, security lighting of streets, gardens and public transportation stops, street signaling, water purification and desalination, railway signaling.
- Agriculture: Water-pumping installations, microdrip irrigation systems, livestock watering and management, electric fencing, automatic feeders.
- Health: Medical facilities in rural areas, refrigeration (very useful in developing countries for the conservation of vaccines and blood), emergency power for clinics, water quality and environmental data monitoring.

2.6 COMPONENTS OF STANDALONE PV SYSTEM

A standalone PV system consists of PV cells, modules or arrays depending on the size of the system. The DC output of PV cells on solar irradiation is used to charge batteries via the charge controller which is connected to DC powered devices. The battery stores energy when the power supplied by the PV modules exceeds load demand and releases it when the PV supply is not sufficient. A charge controller or battery regulator is connected between the solar arrays and the batteries to ensure that the maximum output of solar arrays is used to charge the batteries without overcharging or damaging them. It also ensures a long working life of the battery in order to maintain the system efficiency. The load for standalone systems can be both DC and AC. For AC-powered systems, the DC output of the batteries is converted to AC via an inverter. Standalone PV systems also include balance of system components that include wiring, over-current surge protection and disconnect devices as well as power conditioning equipment. Typical standalone PV system components are:

- Solar source or solar radiation
- PV cells
- PV modules and arrays
- Solar charge controller
- Battery bank
- Inverter
- DC disconnect (optional)
- Backup generator (optional)

2.6.1   Solar Source or Solar Radiation

The source of solar energy is the sun which is located at 1.5×10^{11} m (approximately 92,955,807 miles) from Earth and has a diameter of 1.3×10^9 m (approximately 864, 327.328 miles). Solar energy is created at the core of the sun when hydrogen atoms are fused into helium by nuclear fusion, releasing a huge amount of energy in the process. It is estimated that the sun produces 100 billion MW of power, but only 200 million MW reaches the earth. Some of this energy is released in the form of photons (light) which is an electromagnetic energy that radiates from the sun's surface to the space and down to the earth. Edmund Becquerel, a French scientist, discovered in 1839 that certain materials would give off a tiny amount of electricity when irradiated by this sunlight. This phenomena of converting light into electricity is called the photoelectric effect.

Solar energy is the most important source of regenerative energy which is also the source of water power, wind, and biomass. The total power radiated into space from the sun is about 3.86×10^{26} W. Most of this radiation, as shown in Fig. 2.4, is in the visible and infrared part of the electromagnetic spectrum, with less than 1% emitted in the radio, UV and X ray spectral bands.

Since the distance of the sun from the Earth is 150 million km, only 0.000000045% of the radiated power is intercepted by our planet which amounts to roughly 174,000 terawatt (TW). This amount of energy from the sun provides more energy in 1 hour than the current needs of energy for the whole world for 1 year, as that global average power consumption totals roughly 17 TW. Sunlight strikes the top of the earth's atmosphere with an average intensity of 1366 watts per square meter (W/m^2) which is known as the solar constant that varies by $\pm 3\%$ because of the earth's slightly elliptical orbit around the sun. On the surface of the earth on a clear day, at noon, the direct beam radiation will be approximately $1\,kW/m^2$ for many locations.

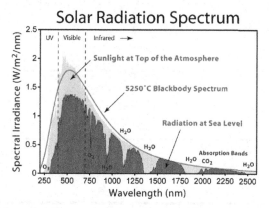

FIGURE 2.4   Solar radiation spectrum. *From Robert A. Rohde, Global warming art. Retrieved from: <http://www.globalwarmingart.com/wiki/file:solar_Spectrum_png>.*

2.6.1.1 Characterization of Solar Radiation

The two most common methods of characterizing solar radiation are solar irradiance and solar insolation. Solar irradiance is an instantaneous power density at a certain location and is measured in units of kW/m^2. It varies throughout the day between $0 kW/m^2$ at night to maximum of $1 kW/m^2$ during day. The solar irradiation also depends on the season, location, and weather.

Solar insolation is a measure of solar radiation energy received at a particular location during a specified time period. It is commonly expressed as average irradiance in kilowatt-hour per square meter per day (kWh/m^2 per day) (or hours/day). In fact solar insolation is the instantaneous solar irradiance averaged over a given time period or typically over the period of a single day. Solar insolation is commonly used for simple PV system design while the solar irradiance is used to calculate system performance of more complicated PV systems at each point in the day.

2.6.1.2 Variation of Solar Energy With Time and Location

The availability of the sun's energy varies depending on the location, season, and time due to the relative motion of sun. However, the biggest factors affecting the available energy are cloud cover and other meteorological conditions which vary with location and time. Generally, more sunlight reaches the earth during midday than during the morning or late afternoon. This is because at midday, the path of the sun rays through the earth's atmosphere is shorter and more radiation reaches the earth's surface. Different part of the world receive different amounts of sunlight and will have more sun hours per day than in other parts, depending on their geographical location. The solar insolation map given in Fig. 2.5 shows the world annual average solar irradiance in kWh/m^2 and is a main tool to consult radiation levels in different part of the world.

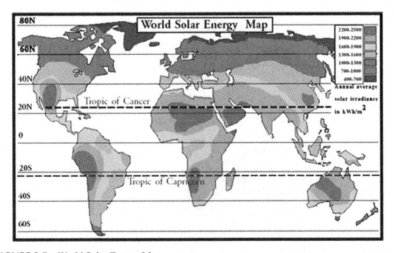

FIGURE 2.5 World Solar Energy Map.

The world solar energy map shown in Fig. 2.5 shows that the availability of solar energy is the highest in regions closer to the equator (Zero degree latitude) which makes these regions a good candidate to use solar energy as a viable source. The availability of solar energy also depends on the seasons and time of the year because of the tilt as the earth revolves around the sun. It is due to this tilt that different locations on earth receive different amounts of solar radiation at different times of the year. In northern Europe, the amount of solar energy arriving at the earth's surface varies over the year from an average of about $0.8\,kWh/m^2$ per day during Winter to more than $4\,kW/m^2$ per day during Summer. The flux changes from place to place and some parts of earth receives much higher annual solar irradiance than others. For example the Red Sea area receives the highest mean irradiance of $300\,W/m^2$ due to its location and climate. Typical values for Australia amount to about $200\,W/m^2$, United States has $185\,W/m^2$ and United Kingdom has $105\,W/m^2$.

2.6.1.3   Dependence of Solar Radiation on PV System Design

In a PV system design, it is important to take into account the seasonal and geographical differences in solar irradiation at a particular time. This can be achieved by making use of measured solar radiation data which is available at a number of locations throughout the world and can be accessed through websites of national governments for most countries. The solar radiation data for the world is also available from the World Radiation Data Center (WRDC) which operates under the auspices of the World Metrological Organization (WMO), located in St. Petersburg, Russia. WMO has been archiving data from over 500 stations and operates a website (http://wrdc-mgo.nrel.gov) in collaboration with National Renewable Energy Laboratory (NREL), located at Golden, Colorado, United States.

NREL also publishes a range of solar maps (http://www.nrel.gov/gis/solar. html) for solar radiation and photovoltaic which provides daily total solar resource information on grid cells of 40 km × 40 km in size. The insolation values shown in these maps represent the resource available to a flat plate collector, such as a PV panel oriented south at an angle from horizontal equal to the latitude of the collector location as needed by the PV system installation. Other orientations are also available. The website also provides useful assessment for US solar energy by displaying average daily direct solar energy incident on a perpendicular surface tracking the sun path and is shown in Fig. 2.6.

Photovoltaic solar resources of the Unites States in Fig. 2.6 shows that the desert in the south west of Unites States consistently receives $6.5–8.5\,kWh/m^2$ per day solar radiation while the southeast receives only about $4.0\,kWh/m^2$ per day. This difference is mainly due to cloud cover and humidity in the Southeast. The solar radiation received by much of Northeast is about $3.0\,kWh/m^2$ per day or less. This means that although the Southwest region of Unites States is most favored region because of high solar radiation, other parts of Unites States are also viable for solar energy systems.

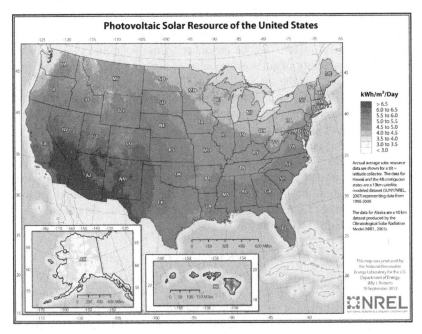

FIGURE 2.6 Photovoltaic resources of the United States.

2.6.2 Photovoltaic Cells

Standalone PV systems use solar arrays consisting of PV cells in order to convert solar radiation into electricity. PV cells are building blocks of the solar module, which in turn are building blocks of PV arrays. The Photovoltaic cell is a thin disc or film of semiconductor material which generates electricity when exposed to sunlight due to the photoelectric effect. The photoelectric effect was first observed in 1839 by Edmund Bequerel, a French Physicist, who found that certain materials would produce small amounts of electricity when exposed to sunlight. The first solar cell was built by Fritts in 1883, who coated the semiconductor selenium with an extremely thin layer of gold to form the junctions (1% efficient). The modern age of solar power technology started in 1954 when Chapin, Fuller, and Pearson of Bell Laboratories discovered that silicon doped with certain impurities was able to generate electricity for satellites. This device originally known as the solar battery is currently called the solar cell and it exploits the principle of the P-N junction. Initially the energy conversion efficiency of the cell was 6%, reaching 11% by 1957 and 14% by 1960. The PV cell is a nonmechanical device usually made from silicon which creates an electron imbalance across the cell and produces direct current as a result of incident sunlight. There are also nonsemiconductor solar cells which are being developed. Three generations of development of PV cells are briefly described in the following paragraphs.

2.6.2.1   First Generation PV Cells

Monocrystalline PV cells belonging to the first generation of PV cells are made from a single crystal ingot of highly pure molten silicon usually crystalline silicon (c-Si). These PV cells were developed in the 1950s as first generation solar cells. Monocrystalline PV cells use wafers, of about 0.3 mm thick, sawn from an Si ingot of single crystal silicon made by using Czochralski process that was discovered in 1916. It is a method of crystal growth used to obtain single crystals of semiconductors, metals, and salts with the most important application in the growth of large cylindrical ingots of single-crystal silicon. Monocrystalline cells are more expensive to manufacture and typically have a slightly higher efficiency of 15–22% than do conventional polycrystalline cells. Shockley and Queisser, in 1961, calculated the maximum thermodynamic efficiency for the conversion of unconcentrated irradiance into electrical free energy in the radiative limit to be 31%. Monocrystalline PV cells are very efficient, approaching their theoretical efficiency in a semiconductor with bandgaps ranging from about 1.25 to 1.45 eV but are very expensive due to the manufacturing processes used to make them. Mass produced solar cells are less efficient and achieve only 10% efficiency.

Polycrystalline PV cells belong to the second type of first generation solar cells and are made from a slice cut from a block of silicon. These PV cells contain multiple silicon crystals which makes it easier to produce wafers in molds from multiple silicon crystal than from a single crystal, making it less expensive. Polycrystalline PV cells are slightly less efficient than monocrystalline but have improved efficiency over the last few years. They are now competitive with monocrystalline in efficiency but are less costly. Polycrystalline (multicrystaline) is made by sawing a cast square ingot block of silicon first into bars and then into wafers. In this process molten multifacet crystalline silicon is first poured into a large molding container and carefully cooled and solidified. Polycrystalline solar cells typically have a slightly lower efficiency of 13–15% resulting in larger individual cells and thus typically a slightly larger module.

2.6.2.2   Second Generation PV Cells

To reduce the high cost of crystalline silicon which makes up to 40–50% of the cost of finished product, in the 1990s and early 2000s the industry developed the second generation solar cells based on thin film semiconductors. These cells can be made from thin film of about 1 μm silicon known as amorphous silicon cell (a-Si) which is the noncrystalline form of silicon. Amorphous silicon has a larger absorption coefficient because of quasi-direct bandgap and hence needs thinner absorbing layers and less material cost than crystalline silicon which absorbs less light near its band edge. These cells are easy to manufacture in one step at very low temperature by depositing a thin film or layer of semiconducting material on a substrate such as glass, metal, or plastic. These cells with bandgap of ~1.7 eV can be doped in a fashion similar to c-Si, to form p–n-type semiconductor junctions and often used to produce large-area photovoltaic solar cells. Amorphous silicon cells, however, suffer from light induced instability that causes the cell efficiency to degrade with time.

Copper indium gallium diselenide (CIGS) and cadmium telluride (CdTe) are other nonsilicon thin film technologies that are based on compound semi-conductors. CIGS or $Cu(In_xGa_{1-x})Se_2$ (tunable bandgap of 1.04 eV to about 1.68 eV) was developed in the 1980s as an alternative to amorphous silicon to improve on its deficiencies. It is formed by depositing thin layers of cop-per, indium, gallium, and selenium on either glass or stainless steel substrate with a complex heterojunction model. CIGS (bandgap of~1.38 eV) has a high absorption quality and allows 99% of the available light to be absorbed in the first micron of material which is applied to energy conversion. In 2014, a CIGS solar cell was reported to achieve more than 20% efficiency at the Center for Solar Energy and Hydrogen Research in Stuttgart, Germany. CIGS, however, degrades rapidly in the presence of moisture.

Cadmium telluride (CdTe), developed in the 1990s, is a direct band mate-rial with a bandgap energy of about 1.45 eV. It has high absorption and is well matched to the solar spectrum for converting solar radiation into electricity using a single junction. A typical CdTe solar cell has a similar heterojunction structure to CIGS consisting of simple p–n heterojunction of p-doped CdTe layer matched with an n-doped cadmium sulfide layer that acts as a window layer. CdTe solar cells are much cheaper than crystalline silicon cells because they use low cost manufacturing technology. In 2014, the Arizona-based firm First Solar reported laboratory efficiency of the new CdTe solar cell to be 21.5% and average commercial module efficiency to be 14.7%. However CdTe con-tains cadmium which is one of the top deadliest and toxic materials known. As a result, strict precautions for safety are required during its usage, manufacturing, and eventual disposal or recycling.

2.6.2.3 Third Generation PV Cells

Third generation solar cells use new materials including silicon, organic materi-als, polymers, conductive plastics, solar dyes, quantum dots, and solar-ink using printing technologies. The goal of the third generation solar cell is to obtain high efficiency and lower per watt cost of the electricity generated. The purpose of using new material like polymers or plastic cells is to use inexpensive, well-known, roll-to-roll technology, similar to that used for printing newspapers. The performance of third generation solar cells is still limited compared to the first and second generation solar cells, but they have great potential in achieving low cost efficient solar cells. Some of these solar cells are already commercialized.

Third generation PV cells also include high performance multi-junction solar cells which are primarily used for space and military applications because of the need for high efficiency irrespective of higher cost of manufacturing.

Multijunction solar cells achieve much higher conversion efficiency by making use of two or more different cells with more than one band gap and more than one junction to generate electricity. These cells are based on a selection of III–V compound semiconductor material where each junction or subcell absorbs and converts sunlight from a specific wavelength of the solar radiation spectrum. Multijunction solar cells consist of a stack of individual single junction in descending order of band gap so that the sunlight strikes the top cell with the highest energy photons. Gallium arsenide, amorphous silicon, and copper indium diselenide have been used in a multijunction cell as one or all of the component cells. In order to reduce the high cost of power production from the multijunction cell, an optical concentrator made of plastic lens or mirror is used to focus sunlight onto a very small piece of costly multijunction cell. The use of this inexpensive concentrator makes the multijunction cell cost effective in industrial applications because a little of the PV material is needed for the cell. In 2014, a new world record of 46% efficiency was achieved in a four-junction cell in which each of its subcell converts 25% of the incoming photons in the wavelength between 300 and 1750 nm into electricity. The work for this multijunction cell was undertaken by Soitec and CEA-Leti of France in collaboration with the Fraunhofer Institute for Solar Energy Systems ISE of Germany.

2.6.2.4   Principle of a PV Cell

The principle of a PV cell is based on the well-known photoelectric effect and semiconductor physics. In semiconductor physics, a photon with energy greater than bandgap energy ($h\nu > E_{gap}$) incident on a semiconductor can excite electrons from the valence band to conduction band, allowing for current to flow. The energy of photon (E) which appears in the form of packets of energy is given by

$$E = h\nu$$

h is Planck constant ($6.626 \cdot 10^{-34} \, W s^2$), and ν is frequency in Hertz.

The maximum current density is given by the flux created by that of incident photons energy. The excess energy which is the difference of incident photon energy and energy gap ($h\nu\text{-}E_{gap}$) is lost in the form of heat or thermalization and is therefore wasted.

A PV cell is essentially a large area p–n junction diode. Energy from the incident photons creates excitation of the electron to the conduction band, leaving behind a hole in the valence band resulting into electron-hole pairs or excitons in the case of organic semiconductors. The electron-hole pairs separate at the p–n junction with electrons and holes diffusing across depletion zone to the p-type and n-type region respectively. This flow of charge carriers due to thermal energy generates a current which is essentially due to electrons in the conduction band that moves through the semiconducting material. The current generated as a result of incident photons on the p–n junction depends directly on the mobility of the carriers in the material and the exposed surface area of the junction. The amount of current generated by photon excitation in a photovoltaic cell at a given temperature is affected first by the intensity of the incident light and second by the wavelength of the incident light. The materials used in photovoltaic cells exhibit a varying sensitivity with respect to the absorption of photons at given wavelengths because of the different spectral responses to incident light. For each semiconductor material there is a cut-off frequency or incident radiation threshold frequency, below which no photovoltaic effect will take place. Above the threshold frequency, the kinetic energy of the photogenerated electron will vary according to the wavelength of the incident radiation without any relation to the change in the intensity of light. The increase in the intensity of light will proportionally increase the rate of photogeneration in the material used for the photovoltaic cell. The light absorbed by a solar cell in real life application is a combination of direct solar radiation, as well as diffused light bounced off surrounding surfaces. As a result, solar cells are often coated with antireflective material to absorb the maximum amount of radiation possible.

2.6.2.5 Characteristics of PV Cells

PV cells can be modeled as a current source in parallel with a diode. When there is no light present to generate any current, the PV cell behaves like a diode. As the intensity of incident light increases, current is generated by the PV cell, as shown in Fig. 2.7.

Equation of the ideal solar cell model, is given by:

$$I = I_{ph} - I_s[e^{V/(mV)_T} - 1]$$

I_{Ph} is photocurrent in amperes
I_S is reverse saturation current in amperes (approximately range $10^{-8}/m^2$)
V is diode voltage
V_T is thermal voltage $= kT/q$ (25.7 mV at 25°C)
m is diode ideality factor $= 1...5 \times V_T$ (-) ($m = 1$ for ideal diode)

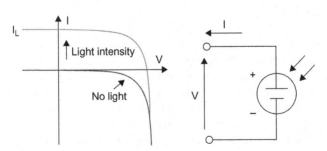

FIGURE 2.7 I-V curve of PV cell and associated Electrical Diagram.

Efficiency (η) of Photovoltaic cell: is defined as the ratio of electrical power (P_{out}) produced by a photovoltaic cell at any instant to the power of the solar input (P_{in}) which is sunlight striking the cell measured in W/m^2. P_{out} can be taken to be P_{MAX} because the solar cell can be operated up to its maximum power output to get the maximum efficiency.

$$\eta = \frac{P_{out}}{P_{in}} \rightarrow \eta_{MAX} = \frac{P_{MAX}}{P_{in}}$$

η_{MAX} (maximum efficiency) found from the light test is an indication of the performance of the device under test. It can be affected by ambient conditions such as temperature, intensity and spectrum of the incident light.

Fill Factor (FF): Fill factor is defined as the ratio of photovoltaic actual power (P_{out}) to the theoretical power (P_T) if both the open circuit voltage and short circuit current were at their maximum. It is a measure of evaluating performance of the cell. Typical FFs range from 0.5 to 0.82. FF is often represented as a percentage.

$$FF = \frac{P_{max}}{P_T} = \frac{I_{mp} * V_{mp}}{I_{sc} * V_{oc}}$$

2.6.2.6   Progress in PV Cells Efficiencies

A PV cell's efficiencies vary from approximately 8% for amorphous silicon based solar cell to approximately 46% for multijunction devices. NREL has been publishing confirmed conversion efficiencies of best research cell from 1976 to present. The conversion efficiencies up to 2015 is shown in Fig. 2.8, and are listed in descending order as given below.

1. Multijunction cell with highest efficiencies
2. Single junction GaAs cells
3. Crystalline silicon cells
4. Thin film technologies
5. Emerging photovoltaic cell technologies with the least efficiency

Best research cell-efficiencies in Fig. 2.8 shows that all conversion efficiencies of PV cell technologies increased with time. The multijunction cell converts 46% of sunlight into electricity which is a new world record. Crystalline based cells have already reached close to the theoretical thermodynamic limit of 31%. The conversion efficiency of thin film also increased but is still lower than the full potential. The efficiencies of emerging dye cells and organic PV cells are currently low but have high potential because of lower cost of manufacturing.

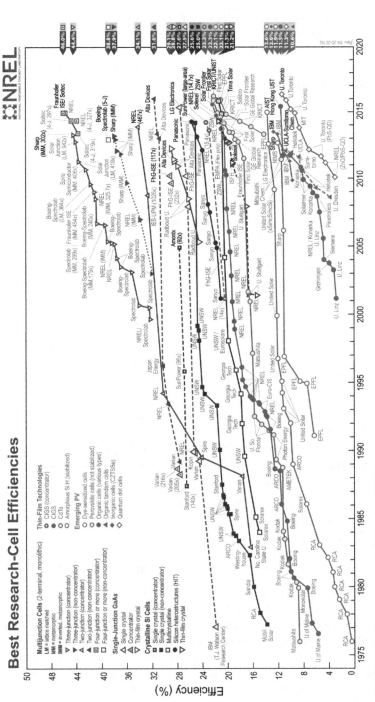

FIGURE 2.8 Best research-cell efficiencies (NREL).

2.6.3   PV Modules and Arrays

A PV cell produces voltages in the range of 0.5–0.6 DC, regardless of its size. The current output of a PV cell varies from 2 to 5 amperes depending on the size and efficiency. The output is also proportional to the intensity of sunlight striking the surface of the cell. Most of the solar panels use rigid crystalline cells but some panels also use thin film cells such as copper indium gallium selenide or cadmium telluride. A small output voltage from a PV cell is not adequate for any commercial applications. As a result, several PV cells are connected together in series to increase the voltage or connected in parallel to increase the current. This arrangement of the interconnection of cells is called a module or is sometimes called a panel. A PV array is the complete power-generating unit, consisting of several modules as shown in Fig. 2.9.

PV cells are usually chosen to be compatible with a 12-V battery. A 12-V battery may require voltages of 15 V or more to charge them. The reduction in PV module voltage due to temperature and other losses can be an obstacle to the charging of PV cells.

Considering a single PV cell produces 0.5 V, most PV modules will use 36 cells connected in series to produce 18 voltage peak output. This will drop to 12–14 V when charging. PV panels are also arranged in modules to produce increments of 12 V such as 24 and 36 V. For 24 V configuration, there will be two 12-V groups of 36 cells (totaling 72 cells) connected in series allowing the module to output 24 V. A typical 12-V module containing 36 cells measures about 25 by 54 inches. The industry standard for residential applications are 60 cell module which typically measures 65 by 39 inches. Commercial panels contain 72 cells and measures 77 by 39 inches. The depth of the solar panels range between 1.4 to 1.8 inches with most of the manufacturers moving toward 1.8 inch depth. The solar panels for residential applications weigh approximately 40 pounds and commercial panels weigh approximately 50 pounds. The size of the thin film solar module is much larger than that of a crystalline based module because of its lower efficiency. However, its weight is lower than the weight of a crystalline-based module.

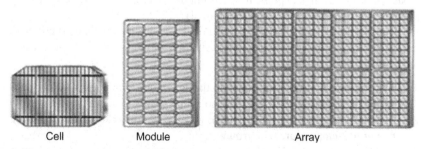

Cell   Module   Array

FIGURE 2.9   Photovoltaic cells, modules and arrays.

PV panels are manufactured in different sizes and can produce power as low as 5 W or as high as 300 W. Solar panels producing less than 50 W are often used for standalone systems such as solar pumps, traffic lights, motorist call boxes, charging batteries for cell phones, solar lights and other small scale appliances in the aftermath of a disaster. Higher power panels producing over 50 W are used for commercial and residential applications including grid-tied systems. There is an increased trend in using large size panels because these systems are cheaper to produce and take less time to integrate as compared to small systems. However, the disadvantages of large systems include increased weight, resistance to wind loads, and problem of packaging, shipping, and handling the panels. PV panels have the advantages of being small but are more expensive in terms of power production per unit watt. These interconnected cells and their electrical connections are encapsulated with tempered glass or clear plastics on the front surface of protective and waterproof material such as plastic and metal on the back surface. The edges are sealed for waterproofing and an outer frame of aluminum is attached to hold everything together. Some thin film modules use flexible backsheet material such as aluminum or stainless steel substrate and plastic to replace glass backsheet material. Such materials can also be designed to increase dielectric strength and reflectance to increase the efficiency of the cell and ensure a long life of the module.

A junction box or wire leads are attached to the back of the module to provide electrical connection to the outside world. The junction box may include a bypass diode which serves as a protection mechanism that allows the panel to continue producing power when one of its cell strings is shaded or damaged. Bypass diodes are passive components that require no maintenance.

2.6.3.1 PV Module Characteristics

The energy conversion capability can be described by a current-voltage (I-V) characteristic which is the basic electrical output profile of a PV device. The I-V characteristic represents the combination of all possible currents and voltages of a given PV device (cell, module or array) which could be operated at a specified condition of incident solar radiation and cell temperature. The plot of I-V curve as shown in Fig. 2.10 is expressed by current in amperes on the y-axis (vertical) and voltage in volts on the x-axis (horizontal).

The I-V curve of a PV device is based on the device under standard test conditions (STC), an industry standard for conditions, under which solar devices are tested as given below:

1. Solar irradiance-1000 W/m^2 refers to the amount of light energy falling on a given area at a given time. This is also called peak sun or one sun.
2. Temperature of the cell 77°F (25°C). It is the temperature of the cell itself and not the temperature of surroundings.
3. Air Mass-1.5 (AM-1.5). It refers to the amount of light that has to pass through earth's atmosphere before it hits earth' surface. AM-1.5 is minimized

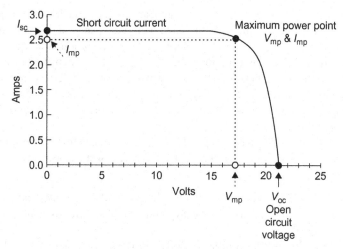

FIGURE 2.10   PV module I-V characteristics.

at noon when the sun is above the PV system and increases as the sun goes farther from the reference point at a certain angle. The basic I-V curve parameters include four attributes as given below:

Open circuit voltage (V_{oc}) is the maximum voltage available from a PV device under infinite load which occurs at no current. The power output is also zero because of zero current output. In case of a PV cell, open circuit voltage corresponds to the amount of forward bias due to the bias of the solar cell junction because of the incident light generated current.

Short circuit current (I_{sc}) is the maximum current of the PV device under no load or short circuit condition and no voltage output. The power output is also zero because of zero voltage at the output.

Maximum power point (P_{mp}) is the operating point on I-V curve where the product of current and voltage is at maximum power.

Maximum power current (I_{mp}) is the operating current on an I-V curve where the output power is at maximum.

Maximum power voltage (V_{mp}) is the operating voltage on an I-V curve where the power output is at maximum.

Maximum power in watts $= (V_{mp}) \times (I_{mp})$

$$V_{mp} = \text{maximum power voltage in volts}$$

$$I_{mp} = \text{maximum power current in amperes}$$

2.6.3.2   Quality Assurance and Safety Requirements of PV Modules

Quality assurance of solar modules is very important for failure-free operations, longer life and efficient power generation. In low-quality standalone PV modules, problems of microcracks can appear due to the lack of strict quality control. This requires a simple and reliable method to evaluate the performance of the solar module for failure-free operation during the production process and also after the module has been installed. One relatively expensive but reliable method to evaluate the performance is by way of thermal imaging camera. This method shows a clear image of any abnormality of large areas within a short timeframe and can be used to scan installed solar modules before and during normal operation.

Photovoltaic modules are designed and expected to last for several years and are subject to many standards, including PV safety testing services, as given below:

- ANSI/UL 1703: American National Standards Institute/Underwriters Laboratories, Inc. Standard for Safety for Flat-Plate Photovoltaic Modules and Panels.
- ANSI/UL, 1703-2012(ANSI/UL, 2002): Refers to version of UL 1703 with a revision date of May 8, 2012 or earlier.
- ANSI/UL 1703-2013: Refers to new edition of UL 1703 that incorporate the new fire classification.
- IEC 61730: PV Safety Standard for Europe and Asia.
- NFPA 70E (2012 edition) help companies and employees avoid workplace injuries and fatalities due to shock, electrocution, arc flash, and arc blast.
- EC 61215: International test standard for crystalline silicon terrestrial PV modules
- IEC 61646: International test standard for thin-film terrestrial PV modules
- IEC 62108: International test for standard for concentrator modules/assemblies

2.6.3.3 PV Arrays

A photovoltaic array consists of a number of individual PV modules that have been wired together in series or in parallel to increase the voltage or amperage. It is a complete power generating unit which varies from a few hundred watts to hundreds of kilowatts. However, larger systems are often divided into several electrically independent subarrays each feeding into their power conditioning systems. In standalone systems, the array must be sized to produce enough power to meet a specific load during the period with the greatest load and lowest solar radiation incident at that region. In addition, some excess power is also required to account for voltage drop, inefficiencies in battery charging and other losses in the system. PV arrays can be mounted on roofs, poles, or racks depending on the applications. Roofs are convenient locations for grid-tied systems in a built-up area while pole mounting may be used in the remote areas or in standalone systems. The advantages of roof-mounted systems include better sun exposure, better protection for the modules from physical damage and saves space on the ground for other usage. Its disadvantages include hazards caused by lifting all the modules and accessories to the roof, susceptibility to the leakage at the attachment points and difficulty in case of repair and maintenance.

The performance of PV modules and arrays are rated for their maximum DC power output according to well defined set of conditions known as Standard Test Conditions (STC). These conditions are defined by the temperature of a module cell 25°C (77°F), the intensity of incident solar radiation ($1 \, kW/m^2$) and spectral distribution of the light (AM 1.5). As these conditions are not always met in the operation of PV modules and arrays in the field, actual performance is usually 85–90% of the STC rating.

2.6.4   Charge Controller

The charge controller, which is connected between the PV generator and the battery (Fig. 2.11), is the most important component in the PV standalone systems with battery storage. Its purpose is to keep the system batteries charged and safe for a long time. The main function of the charge controller is to charge a battery without permitting overcharge and at the same time, preventing reverse current flow when there is no sun. Charge controllers, although not needed for smaller systems, are required for systems with optimized undersized battery storage to reduce initial cost, and systems which have user intervention.

Essentially, the charge controller is used to protect battery overcharge or undercharge and for automatically connecting and disconnecting an electrical load from sunset to sunrise.

Overcharging often occurs during the summer when the photovoltaic array is operating under good or excellent weather conditions, resulting into generation of energy exceeding the electrical load demand. In the absence of a charge controller, the current from the array proportional to the amount of sun radiation will flow into a battery, even if the battery does not need charging. In the case of the fully-charged battery, the unregulated charging will cause the battery voltage to rise to an exceedingly high level, resulting in to internal heat, electrolyte loss, severe gassing and grid corrosion. Overcharging of battery can reduce battery life span, battery performance, and may pose a safety risk. A charge controller prevents such a battery overcharge by limiting or interrupting current flow from the photovoltaic array to the battery when it becomes fully charged.

Overdischarging takes place during periods of low solar radiation and excessive electrical load usage which causes insufficient energy from the photovoltaic array to keep the battery fully recharged. In the case of a deeply discharged

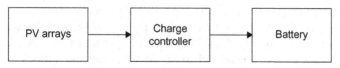

FIGURE 2.11   Block diagram of a charge controller arrangement.

battery the chemical reaction in the battery takes place close to the grids, weakening the bond between active materials and the grids. In the case of repeatedly overdischarge of batteries, loss of battery life and loss of capacity occurs. Overdischarge protection is accomplished by disconnecting the system loads when the battery reaches a low voltage. In a charge controller this is achieved by open-circuiting the connection between the battery and electrical load when the battery reaches a preset or adjustable low voltage load disconnect set point. The loads are again reconnected once the battery reaches to a certain level.

2.6.4.1 Type of Charge Controllers

A basic charge controller monitors the battery voltage and stops charging, when the voltage rises to a certain level, and starts again when the voltage drops below a certain level. As a result one way of characterizing the charge controller is by the way they regulate charging current to a battery. In a shunt charge controller, charging current to a battery is limited by short-circuiting the PV array without any harm, unlike the batteries. A shunt element inside the charge controller is used to short-circuit the PV array which moves the array's operating point on the I-V curve near the short-circuit condition and limits the power output. A blocking diode in series between the battery and the shunt element is used to prevent the short circuiting of the battery. In a series charge controller, charging the current to the battery is limited by open circuiting the PV arrays. A switching element inside the charge controller opens when the battery reaches a full state of charge. This will move the array's operating point on the I-V curve to the open circuit condition and limit the power output. The switching methods to regulate charge are based on certain algorithms which are selected in conjunction with the step-points to optimize battery charging and overall system performance. Most charge controllers use solid state switching element like a transistor which passes current in one direction and prevents reverse current. Most of the modern charged controllers work in a three stage charge cycle as given below:

1. Bulk: The controller sets a point for all three stages for the proper selection of each battery type and voltage. In this phase, voltage rises gradually to around 14.4–14.6 V for a 12-V system. During this time the battery draws maximum current resulting in a state of charge of about 80–90%. Once the voltage rises to bulk level, the absorption stage starts.
2. Absorption: In the absorption phase, voltage is maintained constant at bulk level for specified time while the current gradually tapers off (reduces) as the batteries charge up. Once the battery is full, the final stage starts. The absorption charging period can be preset or adjustable, and is usually 1–3 hours.
3. Float: After the absorption time elapses, the voltage gradually lowers to the float level to around 13.4–13.7 V for a 12-V system, which is nearly 100% state of the charge. The float charge must not exceed the self-discharge rate or the battery will be overcharged. At this phase the battery draws small

maintenance current until the next cycle starts. Float charging is also referred to as "trickle charging." It is critical to maintain the optimal float levels of the voltage because the gas produced by the battery tends to rise to an explosive level.

2.6.4.2   1 or 2 Stage or On/Off Controllers

A 1 or 2 stage controller is also referred to single or dual voltage charge/diversion controller. This depends on relay or shunt transistors acting as switches to control the charging voltage in one or two steps. It is essentially a simple on/off type of controller which monitors the battery voltage and turns off the power from the PV panels when the battery voltage rises to a certain level. Whether it is designated 1 or 2 stage control, it refers to a number of different steps in a charging routine of the battery. A 1 stage charge controller means that the solar panel is directly connected to a battery without any regulator which can lead to undesired overcharging of batteries. A 2 stage charge controller uses two steps charging routine, the first being the bulk and second the float stage. These systems are capable of operating from a few amps to hundreds of amperes capacity depending on the limitation of switching relays. These charge controllers are best used in applications where the load is always on and it also prevents the battery from being overcharged. However it suffers from an inability to fully charge the battery quickly and does not offer any battery maintenance or equalize charging. These systems are reliable and inexpensive as they only use few components. Applications of 1 or 2 stage controllers are limited compared to more advanced Pulse width modulation (PWM) and maximum power point tracking (MPPT) control charger, discussed below.

2.6.5   PWM Three Stage Controller

Pulse width modulation (PWM) charge controllers are solid state controllers which are based on more advanced three step charging algorithm than the 1 or 2 stage charge controller. It uses a semiconductor switching element between the PV array and the battery which is switched on/off by PWM at a variable frequency determined by the variable duty cycle to maintain the battery at or very close to the voltage regulation setpoint. In other words, when the charging voltage of a battery reaches the regulation setpoint, the charging algorithm slowly reduces the charging current to avoid heating and gassing of the battery, but continues to return the maximum amount of energy to the battery in the shortest time. This type of trickle charging is ideal for PV arrays where excess energy is produced for days and weeks and very little of it is consumed. Unlike an on/off controller, the PWM controller works by reducing current from the PV array according to the battery's condition and recharging requirement rather than suddenly cutting off the power to minimize battery overcharging. This is achieved by electronically controlling the speed of switching element of the PWM controller which breaks the PV array current into pulses at some constant frequency, and varies the width and time of the pulses to regulate the amount of charge flowing into the battery. The use of pulses in the PWM charge controller is also good for the batteries as it mixes the electrolyte cleaning of the lead plates and prevents sulfation. PWM charge controller maintains battery capacities of 90–95% compared to 55–60% of on/off regulated state of the charge and has the ability to recover lost battery capacity. This can equalize drifting battery cells, automatically adjust battery aging, increase the charge acceptance of the battery and selfregulate the voltage drops and temperature effects in solar systems. The greatest advantages of PWM controllers are that they have been around for many years, are cheaper, and are available in a wider variety of sizes for different applications.

The PWM charge controller is a good low-cost solution for small systems only, when solar cell temperature is moderate to high (between 45°C and 75°C). Drawbacks of using a PWM controller are that when sizing a system the controller must match the voltage of the battery bank, and PWM controllers are normally limited to 60 amps maximum. Most module manufacturers have switched to a 60-cell design, resulting in modules in the 200–300 W range with a maximum power point of 25–35 V.

2.6.5.1 Maximum Power Point Tracking Charge Controller

In a PWM charge controller, if the controller receives more than 18 V from a PV panel in a 12 V system, the excess voltage is usually wasted and not used. However in case of MPPT, the controller will use that extra 18 V DC and convert it into higher current which will result into faster charge time for the charging battery. This results in higher efficiency, higher power, and better overall battery management than PWM controllers. MPPT, unlike PWM using switching algorithm, continuously adjusts the load on a PV device under changing conditions to keep it operating at its maximum power point. It checks the output of the PV array and compares it to the battery voltage. It then calculates the best power that the PV array can produce to charge the battery. MPPT charge controller takes this voltage and converts it to the best voltage to get maximum current into the battery. MPPT's power varies and depends on solar radiation, ambient and solar cell temperatures. In other words, an MPPT charge controller takes into account the fact that PV arrays occasionally produce variable output voltage due to weather conditions, and can automatically match the voltage they produce to the voltage of the battery in order to maximize charging efficiency. The voltage at which a PV array can produce maximum power is called the "maximum power point."

MPPT not only performs the function of a basic PV controller but also operates as a DC voltage converter which converts the voltage of PV arrays into that required by the batteries without any considerable loss. This DC converter changes DC input from PV arrays to AC and then converts it back to DC exactly matching it to the battery voltage. In this conversion, a Buck converter is used for step down voltage and a Booster converter is used for step-up voltage. If the battery voltage is equal or less than 48 V, it is normal to choose a Buck converter while for battery voltages greater than 48 V, a Booster converter is selected.

MPPT, when compared to PWM, has a higher system efficiency and can be used for larger PV arrays with higher output power. Most modern MPTTs are around 93–97% efficient in the conversion and get a typical gain of 20–45% power increase in Winter and 10–15% in Summer over a PWM controller. Most MPPT controllers have the capacity to handle up to 80 amps. MPPT is most effective in low temperature including cloudy or hazy days with insufficient sunlight. It can also extract more current and charges the battery when it is deeply discharged with a lower state of charge. The MPPT controllers, however are more complex and much more expensive than the PWM controllers.

2.6.6   Battery Storage for Standalone PV System

Most PV standalone systems, in the aftermath of disaster and in remote areas, use rechargeable batteries except when a solar power water pumping system is operated. The function of the battery is to store electrical energy during the day when it is produced by the PV array and supply it to the electrical loads when needed at night time. The storage is also needed during long period of cloudy weather because the consumption does not coincide with energy generation by the solar arrays. Batteries in a PV standalone system also provides stable voltages and currents to electrical loads by suppressing transients and peak operating currents. The features of the batteries used in the aftermath of disaster and remote areas are:

- Minimal or low maintenance requirements especially for remote areas
- Longer life and recyclable
- Higher price/performance ratio
- Higher capacity and cycle life
- Low self-discharge rate and higher energy efficiency
- Easily chargeable with small currents
- Easy to handle and install
- Optimum size and space requirement
- Protection against health and environmental hazards
- Vibration resistant during transportation

2.6.6.1   Types of Storage Batteries for Standalone PV Systems

Batteries for PV standalone systems are subject to frequent charging and discharging process. As a result they require a type of a secondary battery which can store and deliver electrical energy and can be charged by passing current in an opposite direction to the discharge current. The storage capacities in the PV standalone systems range between 0.1 and 100 kWh, although a few systems in the MWh are also implemented. For large standalone PV systems, the most common type of true deep cycle battery is the deep cycle rechargeable lead–acid batteries which are capable of handling the constant charging and discharging.

For remote applications gel type lead–acid batteries are used because of the maintenance-free operation requirement.

In small and portable applications, rechargeable nickel cadmium, or Ni-Metal hydride and lithium ion batteries are used. The applications of these batteries include radios, laptops, cell phones, solar flash lights, small medical appliances, cordless power tools, cellular and cordless phones, digital cameras, two-way radios, camcorders, and remote control toys.

2.6.6.2 Lead–Acid Batteries

A lead–acid battery consists of positive electrodes (anode) made of lead oxide (PbO_2) and a negative electrode (cathode) made of porous lead (Pb). Both the electrodes are suspended in a solution of sulfuric acid (H_2SO_4) in water which acts as electrolyte. The chemical reaction during discharge and recharge is normally written as seen here.

Negative electrode reaction during discharging:

$$PbO_2 + HSO^-_4 + 3H + 2e \rightarrow PbSO_4 + 2H_2O$$

Positive electrode reaction during charging:

$$Pb + HSO^-_4 \rightarrow PbSO_4 + H^+ + 2e^-$$

During the discharge operation, acid is consumed and the process is driven by the conduction of electrons from the negative electrode to the positive electrode through an external circuit. During the charging operation, water is consumed and the process is driven by the forcible removal of electrons from the positive electrode and their introduction to the negative electrode. This chemical reaction produces an electrical battery as electricity is the flow of electrons. This happens when any load is connected across the battery with a wire through which generated electrons can be transported. In a typical lead–acid battery, the voltage is approximately 2 V per cell regardless of size. Lead–acid batteries are less expensive and are available in many storage capacities ranging from 10 Ah to over 1000 Ah. The most suitable type of lead–acid battery for use in PV system is the deep discharge battery. This type of battery is designed to provide small amounts of power continuously over long periods and a discharge up to 80% of the total battery charge without damaging it. These batteries can be flooded (wet) type, or sealed as described below.

2.6.6.3 Flooded Lead–Acid Batteries

The flooded (wet) type of lead–acid batteries contains a liquid electrolyte of sulfuric acid and water solution and the electrodes are fully immersed in the electrolyte. During discharge operation, hydrogen and oxygen gases are produced from water and pass through the vents of the battery. This necessitates the batteries to be initially filled with electrolytes and the acid level has to be regularly checked and topped up with distilled water whenever necessary. This requires maintenance and it is not suitable for remote applications. It can also be hazardous.

2.6.6.4   Sealed Lead–Acid Batteries

In order to achieve a maintenance-free lead–acid battery, a sealed battery was developed in 1970s by adding a valve to control the venting of gases during the charging and discharging operation. The purpose of the valve is to act as a safety vent when gases build up during overcharge or discharge. The sealed lead–acid battery facilitates to combine hydrogen and oxygen during charging process and to make water, which prevents the drying out of the battery. It is valve-regulated lead–acid battery with no opening for adding acid. Sealed batteries are also called captive electrolyte batteries. These batteries make a form of immobilized electrolyte which has less electrolytes, hence fewer problems compared to the flooded electrolyte batteries. As these batteries are sealed and liquid tight, they can be installed and operated in any position even sideways. These batteries are used in situations where hydrogen gassing cannot be tolerated, or battery has to move during transportation. There are two types of sealed batteries: gel electrolyte batteries and absorbed glass matt (AGM) batteries.

2.6.6.4.1   Gel Electrolyte Battery

Gel cell battery makes use of a mixture of silicon oxide and sulfuric acid (electrolyte) to immobilize the electrolyte. This mixture when added to the battery as a warm liquid turns into a gel as it cools.

After some time, the gel gradually dries out creating small cracks and voids which provides paths for gases to move around and recombine. These gases, hydrogen and oxygen, produced as a result of chemical reaction at the positive and negative plates, combine to form water. This process eliminates the need for water addition. However, excessive gases escape through the pressure relief vents. This results into drying and shrinking of the gel, causing it to lose contact with the plates and reducing capacity. This type of lead–acid batteries is suitable for standalone PV systems in the remote areas because of less maintenance, easy transportation and no need for water addition. Gel-based sealed batteries, although less hazardous and easy to maintain, can only be used in limited conditions because of limited deep discharges.

2.6.6.4.2   Absorption Glass Mat Electrolyte Battery

The electrolyte (sulfuric acid) in this cell is immobilized by using a thin absorbent fiber glass mat between the lead plates. The fiber glass mat with small pores absorbs the electrolyte acid and immobilizes it while still keeping the acid available to the plates. The fiber glass mat also acts as a separator between the plates. Its construction arrangement allows the hydrogen and oxygen to move around and recombine to make water. The charging rate of AGM is five times faster than that of the flooded battery. Its life cycle is better than the flooded battery. The cost of AGM is more than the flooded battery but is less than the cost of the gel battery. AGM works well as a midrange battery with the capacities of 30–100 Ah. Its limitations include sensitivity to overcharging, gradual decline and low specific energy. AGM is less environmentally friendly than the flooded battery. It is sensitive to overcharging and must be stored in charged condition.

2.6.6.5 Nickel–Cadmium Battery

A nickel–cadmium (Ni–Cd) battery is an alkaline battery consisting of positive electrode made of nickel oxyhydroxide (NiOOH) and negative electrode made of porous cadmium (Cd). The electrodes are separated by nylon separators which are immersed in potassium hydroxide (KOH) electrolyte that is placed in a stainless steel casing. The chemical reaction during discharge and recharge is normally written as:

Negative electrode reaction:

$$Cd + 2OH^- \leftrightarrow Cd(OH)_2 + 2e^-$$

Positive electrode reaction:

$$2NiO(OH) + 2H_2O + 2e^- \leftrightarrow 2Ni(OH)_2 + 2OH^-$$

During charging, the reactions reverse. Hence reaction arrows are usually shown as $\longleftrightarrow$.

Ni–Cd batteries have a long life, more temperature tolerance and low maintenance compared with lead–acid batteries. It also has low self-discharge and nonfreezing features to its advantage. However it is expensive, costly to dispose of, and has limited availability and lower efficiency than lead–acid batteries. The output of Ni–Cd remains constant up to its last moments, which makes it difficult to measure the depth of charge. Because of environmentally regulatory rules, cadmium is being replaced by nickel metal hydrides.

2.6.6.6 Nickel–Metal-Hydride

Nickel–metal-hydride (NiMH) is a practical replacement of Ni–Cd where anode is made of NiMH instead of Ni–Cd. Ni–MH uses positive electrodes of nickel oxyhydroxide (NiOOH) and the negative electrodes uses a hydrogen absorbing alloy instead of cadmium. The electrochemistry of the rechargeable Ni–MH battery is given below:

Negative electrode reaction:

$$H_2O + M + e^- \rightleftharpoons OH^- + MH$$

The charge reaction is read from left to right and discharge reaction is read right to left.

Positive electrode reaction:

$$Ni(OH)_2 + OH^- \rightleftharpoons NiO(OH) + H_2O + e^-$$

Nickel oxyhydroxide, NiO (OH), is formed.

Ni–MH has a higher specific energy with fewer toxic metals, less effect on memory and generates high peak power. It also has good deep discharge and is environmentally friendly. However, NiMH is more expensive, has higher self-discharge and has lower efficiency than the lead–acid and Ni–Cd batteries. Ni–MH is available for AA and AAA batteries and is used to power small appliances including medical instruments, cell radios, and other small appliances.

2.6.6.7   Lithium Ion Battery

Lithium ion battery consists of a positive electrode made from lithium-cobalt oxide ($LiCoO_2$), a negative electrode made from carbon (graphite) and an electrolyte. The positive electrode in these relatively new batteries is also made from lithium iron phosphate ($LiFePO_4$) and electrolyte. The other type of positive electrode includes lithium hexafluorophosphate ($LiPF_6$) dissolved in a mixture of ethylene carbonate and dimenthyl carbonate.

A lithium-ion rechargeable battery consists of a spiral structure with 4 layers. A positive electrode activated by cobalt acid lithium, a negative electrode activated by special carbon, and separator are put together in a whirl pattern and stored in the case. It also incorporates a variety of safety protection systems such as a gas discharge valve which helps prevent the battery from exploding by releasing internal gas pressure if it exceeds the design limit.

The lithium-ion battery makes use of lithium cobalt oxide (which has superior cycling properties at high voltages) as the positive electrode and a highly-crystallized specialty carbon as the negative electrode. It uses an organic solvent, optimized for the specialty carbon, as the electrolytic fluid.

$$\text{Positive electrode (anode) } LiCoO_2 \underset{\text{Discharge}}{\overset{\text{Charge}}{\rightleftharpoons}} Li_{1-x}CoO_2 + xLi^+ + xe^-$$

$$\text{Negative Electrode (cathode) } xLi^+ + xe^- + 6C \underset{\text{Discharge}}{\overset{\text{Charge}}{\rightleftharpoons}} Li_xC_6$$

$$\text{Battery as a whole } LiCoO_2 + Li^+ \rightarrow Li_2O + CoO$$

The electrolyte acts like an insulator inhibiting the flow of any electrons through it. During the charging operation, the positive electrode releases some of its lithium ions which move to the negative electrode through the electrolyte and remain there. At the same time electrons flow from positive electrode to negative electrode but through the external path which takes longer to complete the circuit. During the discharging operation, the lithium ion moves back to the positive electrode across the electrolyte and electrons flow from the negative electrode to the positive electrode through the external circuit which produces the energy that powers the battery. Electrons in both these cases flow in the opposite direction to the ions across the external circuit. There is no movement of ions when the battery is fully charged. When all the ions have moved back, the battery is fully discharged and is ready to be charged again. The Li-ion battery family is divided into the following battery types which are named by their cathode oxides.

- Lithium-ion–cobalt or Lithium cobalt ($LiCoO_2$)
- Lithium-ion-manganese or lithium–manganese ($LiMn_2O_4$)
- Lithium-ion-phosphate or lithium–phosphate ($LiFePO_4$)
- Lithium Nickel Manganese Cobalt Oxide ($LiNiMnCoO_2$ or NMC)
- Lithium Iron Phosphate ($LiFePO_4$)
- Lithium Nickel Cobalt Aluminum Oxide ($LiNiCoAlO_2$)
- Lithium Titanate ($Li_4Ti_5O_{12}$).

A Li-ion battery can store 150 Wh of electricity in 1 kg of battery compared to Ni–MH battery which can store between 60 and 100 Wh of electricity in 1 kg of battery. The lead acid battery can store only 25 Wh/kg which is six times more in weight than storing the same amount of electricity in a Li-ion battery. A pack of Li–ion batteries loses only 5% of its charge per month compared to 20% per month for Ni–MH batteries. The energy density of Li-ion batteries is three times than that of lead–acid batteries. Each cell voltage of Li-ion battery has 3.5 V. The use of fewer cells in series for Li-ion battery will give the required voltage than those of Ni–MH and lead–acid batteries resulting in a smaller size battery pack. Li-ion batteries are more efficient, more compact, and easier to maintain. They last longer, and comparatively more sustainable than other batteries.

The Li-ion battery, however suffers from poor cycle life in high temperatures. The internal resistance of Li-ion battery rises with age and has safety concerns if overheated or overcharged.

These batteries are currently more expensive than Ni–MH and lead–acid batteries. The earlier Li-ion batteries suffered from risk of explosion until the current batteries were developed by Sony in 1970.

When selecting battery types, consideration of their characteristics in response to temperature changes must be taken into account. The lead–acid batteries, when compared to Ni–Cd batteries, operate less efficiently in cold temperatures and will require proper ventilation.

2.6.6.8 Battery Hazards in Standalone PV Systems

Most standalone PV systems use storage batteries that release gases as a result of charging process and hence may pose health and safety problems. Large standalone PV systems use deep cycle rechargeable open vent or sealed lead–acid batteries while smaller systems tend to use Ni–MH or lithium-ion batteries. The rechargeable battery is considered to be the most risk-prone component of standalone PV systems. In lead–acid batteries, a battery bank is often a group of 2 or 6-V deep cycle batteries connected together to provide a higher system voltage. High capacity battery banks are usually constructed by using between one and three series-connected strings of 2V lead–acid cells. Designing the bank system around a single series string of batteries is normally the best approach, but for very large systems this may require paralleling of series strings to obtain the required storage capacity. A battery bank is the most volatile part of the PV standalone system which can be a cause of personnel injury, damage to the system(s), and environment. As a result, storage batteries for PV standalone systems present urgent safety challenges and thus additional installation and maintenance requirements should be met. The battery installation must meet the requirements of NEC code, Article 480 plus the additional requirements of Part VIII of Article 690 for storage of batteries. Batteries are potentially dangerous and users should be aware of main hazards they present, such as:

1. Electrolyte hazard: It comprises of dilute sulfuric acid which is corrosive. It may cause severe chemical burn if handled improperly. In addition to eye and foot protection, protective clothing is also important to use when working with batteries.
2. Electrical hazard: Batteries have a high current-generating capability which can be a cause of a severe electric shock if short-circuited. The shorting of the battery terminal can cause a current of several thousand amperes to flow for a few seconds. This can cause severe burns and loss of life even though the voltage is low.
3. Charging or overcharging hazard: During charging, especially overcharging, batteries may produce an explosive mixture of hydrogen and oxygen gases which may result in a battery explosion, leakage, or irreversible damage to the battery. This may also result into damaging the components to which the overcharged battery is connected. The risk of such an explosion can be reduced by ventilating the battery which will prevent the build up of these gases. To further reduce the risk of explosion, potential ignition sources such as circuits which may generate sparks or arcs should also be eliminated from the battery enclosure.
4. Thermal runaway hazard: This is defined as the generation of heat at a rate higher than a battery can dissipate, which may happen as a result of electrolyte dryout or severe overcharging. The elevated temperature because of heat can emit flames, electrolytes, or dangerous fumes and can damage or destroy the cell or the whole battery. In battery packs, the cells are normally located in close proximity to one another, and the high heat of the one failing cell can spread to the next cell, causing a chain reaction that spreads to other cells. This phenomena is often referred to "Thermal Runaway Propagation" which can lead to a fire and explosions. Safety for battery packs is increased by installing dividers to protect the failing cell from

spreading to the surrounding cells. Thermal runaway could also be caused due to temperature differences, shorted cell, and voltage imbalances at the time of manufacture.

Both the lead–acid batteries as well as the lithium-ion batteries are capable of going into thermal runaway. The likelihood of such an event is higher in a lithium-ion battery because it has higher amount of energy in a smaller volume. The thermal runaway in lithium-ion cells can be triggered by several causes such as overheating of the cell above 60°C (140°F), internal short circuit and physical damage resulting into fire or explosion.

2.6.6.9 Battery Safety Precautions

Some important battery safety precautions that should always be undertaken are:

- Maintain the battery according to the manufacturer's safety instructions.
- Recycle the battery when it wears out.
- Meet the NEC safety code and OSHA regulations.
- Battery should not be charged at voltages greater than gassing voltages for a longer duration.
- Locate the batteries in a secured place, possibly in a closet or shed.
- Locate the battery bank close to controller and PV arrays if possible.
- Restrict the placement and replacement of batteries to authorized people only.
- Use warning signs and displays because of dangers of Hydrogen gas and battery acid associated with charging stations. It is a must both for safety and legal reasons. Battery signs should direct and remind the workers how to avoid any mishap or accidents.
- Lead–acid and wet cell batteries that can give off explosive hydrogen when recharging should be located in a well-ventilated space isolated from other electrical components of the system and away from living spaces.
- Allow adequate room for easy access during repair and maintenance.
- For a home-installed standalone PV system, keep a log of whatever you do to the batteries and to other system components. Keeping a logbook of the system's maintenance may be one condition of the system's warranty.
- Remove all jewelry and watchband for wearing watch while working with batteries.
- Wear protective clothing, gloves, rubber boots, and goggles to cover the eyes while working with the batteries.
- Place batteries always on battery racks, trays or enclosures instead of on floor.
- Do not mix different type of battery models, types and ages of batteries.
- Choose maintenance-free (immobile electrolyte) batteries over flooded batteries.
- Maintenance should be regularly carried out about once every 3 months.
- Use a battery management system (BMS).

2.6.6.10   Battery Management System

Due to the many issues and hazards with the rechargeable batteries discussed above, a BMS is vital to ensure safe operation of a standalone PV system. A BMS is defined as an electronic system that manages a rechargeable battery (single cell or battery pack) by monitoring its state, calculating secondary data, reporting that data, protecting the battery, controlling its environment, and/or balancing it. BMS technology has already been used in portable applications such as cellular phones and laptop PCs.

Almost all types of rechargeable batteries can be damaged by excessive voltages (overcharging), excessive heat and/or improper usage. In some cases, it can result in fire and explosion, leading to loss of life and harm to environment. The purpose of BMS is to detect and log problems within the battery system, reporting the potential errors and operating it safely. Its benefits include increased battery lifetime, improved reliability, maintaining the health of all cells to deliver the required power and subsequently a reduction in operating costs of the battery system.

The traditional BMS consists of achieving two functions: monitoring and protecting the battery. The monitoring function includes the measurement of currents, voltages, and temperatures of a cell. The protection function refers to turning the system on or off depending on the values of the measurements. In the event of abnormal or irregular conditions, an alarm might be issued. A more complex BMS includes a charge control function to charge the battery pack, in addition to applying some equalization methods to balance the cell voltage for maximizing battery capacity. A more complex BMS monitors many factors affecting the performance and life of the battery as well as ensuring its safe operation. These BMSs may monitor one cell or multicell battery systems. Some BMSs use computers for advanced monitoring, logging, email alerts and more. The following factors are monitored and controlled by BMSs:

- Voltage: Total battery voltage, or voltages of individual cells
- Current: Battery current in or out of the battery (charge and discharge rate)
- Temperature: Temperature of batteries, or temperature of individual cells
- State of charge or depth of charge: Showing the charge level of battery
- State of health: Overall condition of the battery and cells

Building blocks of BMS can consist of various multicell battery monitor ICs and external components that perform the following functions. External components include contactors/relays to connect the battery pack to the external load or to the charger which are also used to disconnect the battery pack from rest of the system in the case of safety issue or cell failures. In larger systems, BMS communicates with a system controller, charge inverter/charger and other components in the system. Many types of ICs are available for BMS from semiconductor companies like Altera, Philips, Power smart and Texas Instruments. Given below is a list of ICs which are used for BMS application.

- Cell voltage sensors: used for measuring and monitoring cell voltage of each cell within a battery pack which is essential in determining their overall health.
- Temperature sensors: used for each cell and they are used not only for safety conditions but also used to determine whether it is desirable to charge or discharge a battery.
- Fuel gauges or battery current indicators: used to keep track of charge entering and exiting the battery pack.
- Cut off FETs: allow the connection and isolation of the battery pack between the charger and load.
- Cell balancing: compensates the weaker cells by equalizing the charge on all the cells in the chain to extend the overall battery life.
- Microcontrollers: used to manage the information from the sensors and to make decisions with the received information.

2.6.7   Inverter for Standalone PV Systems

An inverter is a device which converts low DC voltages to 120/220 voltages AC and is connected either at the output of PV array or battery bank when AC loads are to be used. Power inverters are available for use on 12-, 24- or 48-V battery bank configurations. The inverter is a major component of a PV systems both for standalone or grid connected systems. In standalone or off-grid systems, inverters are connected to chargeable batteries as the DC power source and operate independently of the PV array. The job of the DC inverter is to provide sufficient AC power of the right voltage and frequency from the DC power provided by the PV array which varies depending on time of the day, temperature, and seasons. In recent years there has been a considerable increase in demand for a number of inverters to be installed on standalone systems which supply power to remote areas, disaster-prone areas, and places with no access to grid. Inverters also enable a user to switch off all electrical current at the time of power failure, which is a useful feature for repairing, maintaining, troubleshooting, or upgrading the PV system.

For any grid-tied PV system, the inverter not only changes the DC from PV array to AC used in electric grid but it also regulates the PV system by making use of a special algorithm known as MPPT (referred to previously in this chapter). A grid-tied PV system often uses a string or central inverter in home and commercial applications which is a large box that is often situated some distance away from the solar array. Microinverters, on the other hand, are small grid-tied inverters that mount on each solar panel, sized to suit individual solar panels rather than a string of solar panels. Solid state inverters use electronic devices used to switch DC power to AC as these devices are easy to repair, need minimal maintenance, and have higher efficiency. There are many types of electronic components such as transistors and thyristors which can perform switching functions. Among the many kinds of transistors, power inverters use metal oxide semiconductor field effect transistor (MOSFET) or insulated gate bipolar transistor (IGBT). MOSFETs operate at lower voltages with higher frequencies of around 800 kHz and lower resistance than IGBT. As a result MOSFETs are generally used for inverter applications in the range of 1–10 kW. IGBT have very low on-state voltage drop and have excellent forward and reverse blocking capabilities but switch at lower speed up to 20 kHz. IGBTs are used in high voltage applications exceeding 100 kW.

Some solid state inverters use thyristors or silicon control rectifiers which can be used as an electronic switching device in the operation of inverter. Thyristors have three leads called a gate, an anode (positive terminal) and a cathode (negative terminal). It can be turned on when a small trigger pulse of current is applied to the gate, and the anode is positive with respect to the cathode. Gating pulse has no effect when the device is reverse biased. These devices behave like a mechanical switch which can only be completely on or completely off. These devices are used in high power applications of several megawatts.

The DC batteries are charged by the PV array which does not directly influence the operation of the inverter. Inverters can also be used to charge the batteries if connected to the utility grid or an AC generator for a totally independent stand-alone solar power system. Small inverters use 12 V DC, but many of the larger ones are designed for 24 or 48 V usage.

Inverters are rated by the total power capacity and characterized by power dependent efficiency and harmonic distortion. As the main function of the inverter is to keep a constant voltage on the AC side and convert the DC input power into the AC output power, inverter efficiency is defined as the ratio of output power to the input power.

$$\eta = P_{AC}/P_{DC}$$

where η is inverter efficiency

P_{AC} = AC power output in Watts
P_{DC} = DC power input in Watts

High quality standalone inverters producing sine wave output have peak efficiencies of about 90%. Some inverters have good surge capacity for starting motors while others have limited surge capacity. In order to maximize energy output, it is important to select an inverter with the following characteristics:

- High efficiency
- Low standby losses
- High surge capacity
- Low harmonic distortion

2.6.7.1 Type of Inverters

There are three types of standalone inverters that differ in output waveform, size, and applications:

- Square wave inverter
- Modified sine wave inverter (modified square wave inverter)
- Sine wave inverter

The square wave inverter produces an alternating current waveform resulting from switching DC between maximum positive and negative values every half period. Its output waveform is a "flattened out" version of sine wave which shifts from negative to positive where it stays for half a cycle, and jumps to full negative and stays there for half a cycle then repeats. This is the simplest and cheapest type of DC to AC inverter and is easy to build. Square wave inverters have higher risk of generating odd harmonics and have no regulation of outgoing voltage. Because of its low power quality and low efficiency, it is not recommended for home applications but has limited applications like powering tools using universal motors.

Modified sine wave inverters (modified square wave or quasi-sine wave) produce an AC in the form of stepped square waveform somewhere between a square wave and a pure sine wave. Its output waveform is more like a square wave but with an extra step. The modified sine wave inverter has better efficiency and fewer harmonics and better voltage regulation than square wave inverters. These inverters are capable of operating a wide variety of AC loads both electronic and household including computers, TV, VCR, satellite receiver and printers. Modified sinewave inverters produce radio frequency noise that can interfere with the working of radio and TVs. These inverters may not operate correctly in some appliances that use motor speed controls and timers. Due to the presence of some non 60 Hz components, modified sine wave inverters can make a clock faster or not run at all. These inverters also introduce harmonic distortion in inductive and audio equipment caused by the harsh clipping in the on off phase changing in voltage. However modified sinewave controllers are much cheaper and less complicated than sinewave inverters. They are the most popular and economical power inverters for many standalone systems and it is recommend to supply automatic power to a normal home using a wide variety of electrical devices.

Sine wave power inverters produce the closest to a pure sine wave of all power inverters which can be used to run on any type of AC equipment. Many true sine wave power inverters are computer controlled and will automatically turn on and off as AC loads vary. These inverters are capable of saving battery power by shutting down when the connected appliances are turned on or off. True sine wave inverters are based on a technique similar to the inverters used to connect it to the utility grid but uses simple circuitry without any protection from or synchronization with grid. They are more expensive than modified sine wave inverters but use less power and run more efficiently. Sine wave power inverters can be used for almost all applications including sensitive audio electronics because its output waveform contains low total harmonic distortion.

2.6.7.2 Installation and Listing of Inverters

Installation of inverters is dictated by the national codes and standards authorities that determine the safety and installation requirement of inverters used in

PV systems. One such organization in United States is National Fire Protection Association (NFPA) which developed the NEC (National Electrical Code), first published in 1897, and is updated every 3 years. NEC is approved as an American national Standard by the American National Standards Institute (ANSI). NFPA is responsible for 380 codes and standards that are designed to minimize the risks and effects of fire by establishing criteria for building, processing, design, service, and installation in the United States as well as in many other countries. NEC addresses nearly all PV power installations including standalone and grid-connected systems with voltages less than 50 V.

Article 690 of the code entitled "Solar photovoltaic system," specifically deals with PV systems and other applicable section of NEC, including over-current protection devices (OCPDs), disconnects, grounding, and utility inter-action. Several of these requirements are based on equipment standards and listing requirements for interactive inverters under UL (undertaker laboratory) 1741 which is a Standard for Inverters, Converters, and Interconnection System Equipment for use with Distributed Energy Resources. Inverters installed in PV systems are also required to conform to certain standards for product listing and certifications and must include a listing mark on their nameplate label. The list should include only those inverters that meet UL 1741 Standards, as tested by a recognized certifying agency. There are several sections within the NEC that require the use of systems to be listed and identified as given below:

- NEC Section 551.32—Recreation vehicles
- NEC section 690.4(d)—PV systems
- NEC Section 690.35(G)—Underground PV source and output circuits for underground use
- NEC Section 694.7(B)—Small wind electric systems
- NEC Section 690.60 and 694.60—Interactive electrical systems for interactive use
- NEC Section 705.4—Utility interactive inverters as interconnected electric power production source interconnection service

2.7   SIZING METHODOLOGIES OF STANDALONE PV SYSTEMS

As the standalone PV system must be able to meet energy requirements every day of the year, it is important to choose each component of the system to fulfill this requirement. As a result the PV system should be sized to determine the number and type of solar panels needed to capture the sun energy, the capacity of the battery for storing the energy during sun and no sun, the types and capacity of charge controllers, converters and electrical cables. The system sizing also needs to consider and decide the number of days of autonomy which is a measure of the reliability of the system. System sizing can be divided into five different steps as given below.

2.7.1   Estimation of Electric Load

The first step in system sizing is to determine the amount of energy that will be consumed on a daily basis. The consumption of energy will determine the amount of electricity that must be produced. The energy consumed or electrical load can be calculated by summing the total DC demand and total AC demand expressed in kilowatts per day. The DC loads, if present should be listed separately from AC loads as energy for AC loads goes through the inverter, resulting in losses that must be accounted for separately. Power consumption for most appliances used as electrical load in PV systems are given in the manufacturer's literature. In the unavailability of this information, it can be estimated by multiplying the operating voltage with maximum current. It can also be found by using meter reading or electric bills for load requirement. A "Kill A Watt" electricity usage monitor can also be used to find out how much power is being drawn at any specific moment, or how much energy is being used over time in watt hours. It is a useful tool to use for finding daily electric load and highly recommended for any battery based system sizing. The load estimate must be as precise as possible to avoid oversizing or undersizing the system. If the design is oversized, money is wasted on excess capacity. If it is undersized, power shortages during operation may result. The average daily load is found by the following formula:

$$\text{Daily load} = \text{wattage} \times \text{time in use} \ (DL = W \times T)$$

$$DL = 100 \text{ watts} \times 2 \text{ hours} = 200 \text{ watt} - \text{hours}$$

Given below is a load table which estimates different kinds of electrical loads with their power, average time of use and average daily energy in watt hours.

Electrical Load	Power (W)	Avg. Daily Time of Use (h)	Avg. Daily Energy (watt-hours)
Lighting	200	6	1200
Refrigerator	300	9.6 (40% duty cycle)	2880
Microwave	1200	0.5	600
Pumps	1000	1	1000
TV and entertainment equipment	400	4	1600
Fans	300	6	1800
Washer	400	0.86 (3 h 2 times per week)	344
Miscellaneous plug loads	200	12	2400
Total all loads	400 W (4 kW)		11,824 Wh (11.8 kWh)

Source: http://www.nabcep.org/wp-content/uploads/2012/08/NABCEP-PV-Installer-Resource-Guide-August-2012-v.5.3.pdf.

In order to save the cost of the consumed power, it is important to examine the estimated electrical load and possibly reduce it especially for home or cabin use. This can be done by adopting the following:

- Identify large and variable loads and possibly eliminate them
- Use large loads during peak sun hours or only in summer
- Find DC alternative models which saves power as inverters are not required
- Replace incandescent fixtures with fluorescent lamps
- Use light-emitting diode lights which are more efficient and last longer

An electric load can also be estimated by using an online tool called "Affordable Solar's Off-Grid Load Estimator" (http://www.affordable-solar. com/Learning-Center/Solar-Tools/Off-Grid-Load-Estimator). This tool allows to input any number of loads into the calculation by inserting their power in watts and time in hours/day used. The online tool will add all loads and give the result in watt-hours that needs to be delivered every day by the system. The power consumption is obtained by multiplying power in watts with time in hours.

2.7.2 Sizing of Battery Bank

Having found the daily power consumption, the next step is to calculate the battery requirements. The number of batteries are determined by knowing how much energy can be stored and compare that to the daily total load and additional power needed in various losses and reserve needed in unexpected losses likely to occur in the PV system. The battery sizing in standalone systems is more critical and should be more accurate than grid-tied systems because the available buffer capacity is quite limited. There are three important considerations in calculating the number of batteries needed in standalone PV systems in addition to three design factors to be considered for sizing the batteries in standalone PV systems.

The first criterion dealing with the amount of storage provided by the battery bank is the "Days of Autonomy." This criterion deals with the number of days a battery bank is expected to provide power to the system without receiving an input from the solar array or when there is little or no sun to recharge the battery. It is a compromise between the time the generator will run in the absence of sun and the added cost of a large battery bank. In case of more days of autonomy there will be larger battery bank and that will be costly. A larger battery bank will need large PV array to recharge the battery bank on a regular basis. Three to five days of autonomy is a good compromise.

The second consideration of sizing batteries is the depth of discharge (DOD) which is defined as how much of the rated capacity of battery has been used or the percentage of energy drained from the battery. The DOD for a battery with a capacity of C100 = 400 Ah which is discharged at 4 A for 20 hours will have DOD = $(4 \times 20/400) \times 100 = 20\%$. (C100 rating means that the battery is designed to be discharged in 100 hours).

Most PV systems use deep cycle lead–acid batteries which can be discharged up to 80%, meaning that they will provide fewer charge/discharge cycles over the life time of the battery, hence shorter life. There should be a balance between the longevity, cost, and problem of replacing batteries. In the system design 50% DOD is normally used, although actual DOD during sunny weather is often less than 20%.

The third factor for considering the sizing of battery bank is the operating temperature. Temperature affects the internal resistance of flooded lead–acid battery and its ability to hold a charge. At lower temperatures below 80°F and high discharge rates, battery capacity is reduced and at higher temperatures it tends to shorten life. The standard temperature for most battery rating is 77°F and batteries perform best in moderate temperatures. A battery that is required to operate continuously at 0°F will provide about 60% of its capacity while the same battery operated continuously in a 95°F (35°C) environment can lose half its expected life.

$$\text{Battery Capacity}\,(\text{Ah}) = \frac{(\text{Total Watts in hours per day used by appliances}) * (\text{Days of autonomy})}{(\text{Depth of discharge}) * (\text{Temp. \& discharge rate derating factor}) * (\text{nominal battery voltage})}$$

The standard value of DOD is 80%. Discharge rate varies with temperature.

After determining battery capacity, the size of batteries can be obtained by consulting the manufacturer's information given in their data sheets that identifies types of batteries with the capacity. It is recommended to use deep cycle batteries for standalone systems. Most of these batteries will withstand daily discharge up to 80% of their rated capacity because of their thicker plates. A deep cycle battery is specifically designed to be discharged to low energy level and will be rapidly recharged day after day. The size of the battery should be large enough to operate the appliance at night and cloudy days. The following example calculates the size of the battery.

Example: A system requires a total battery bank output of 500 Ah. The allowable DOD is 75%, the minimum operating temperature is −10°C (14°F), and the average discharge rate is c/50. Find the required battery-bank rated capacity.

The manufacturer's documentation on battery capacity yields a temperature and discharge-rate derating factor of approximately 80%. Using the above relation for battery capacity.

$$\text{Battery} - \text{bank capacity} = 500/0.75 \times 0.80 = 500/0.6 = 833\ \text{Ah}$$

2.7.3 Sizing of PV Modules

The next design criteria is array sizing which deals with calculating the size and wattage of PV modules. The array sizing in standalone systems must be performed to produce enough electrical energy to meet the load requirement of every day throughout the year. This should ensure that the battery stays charged while accounting for different losses including battery losses of about 20%, module temperature losses of about 12%, possible array shading and a derate multiplier to account for wire losses, module soiling, and production tolerance. The array size is calculated by including daily energy requirement in watt-hour per day, expressed in peak sun hours of the source's location and losses as mentioned. It is obtained by dividing the daily energy requirement by the peak sun hours per day.

Sun hours is defined as equivalent number of hours when solar insolation averages $1 \, kW/m^2$ at a certain location. The number of peak sun hours is numerically identical to the average daily insolation because the peak solar radiation is $1 \, kW/m^2$. Hence a location that receives $8 \, kWh/m^2$ per day can be expressed to have received 8 hours of sun per day at $1 \, kW/m^2$. Sun hours are derived from the total amount of sunlight at a certain location along with other environmental factors including temperature that affects the amount of light and reaches the solar array on the ground. It also depends on additional factors such as orientation of the solar array or how it tracks the sun.

The peak sun hours can be obtained from a free online NREL's National Solar Radiation Data base (http://pvwatts.nrel.gov/version_5_3.php) also known as PVWatts®. It was updated on April 22, 2016.

Another free online tool to estimate the solar electricity production of PV system is available from Solargis (http://solargis.info/doc/free-solar-radiation-maps-GHI) for Europe, Africa and Asia. It calculates the monthly and yearly potential electricity generation in kWh of a photovoltaic system with defined modules tilt and orientation.

$$\text{Array size(kW)} = \frac{\text{Total kWh used by appliances} / \text{day}}{(\text{Solar hours} / \text{day}) * (\text{Efficiency factor } 0.72)}$$

kW hours/day = (kW hours/month) ∗ 1 month/30 days
kW hours/day = (kW hours/year) ∗ 1 year/365 days

Efficiency factor 0.72 is based on the following assumptions

Average solar access = 95%
Inverter efficiency = 96%
DC and AC wiring derate of 0.8 and 0.99
Module soiling rate = 0.95
Module mismatch derate = 0.98
Module temperature derate factor = 0.88
System availability rate = 0.99

$$\text{Efficiency factor} = 0.95 \times .96 \times 0.8 \times 0.99 \times 0.95 \times 0.98$$
$$\times 0.88 \times 0.99 = 0.72$$

Example: A typical household uses 900 kW/month near Los Angles. Solar hours received at this location is 5.62 kW/day. The array size is calculated as:

$$\text{Array size} = \frac{(900 \text{ kW} \div 30 \text{ days per month})}{(5.62 \text{ kWh per day}*0.72)} = 7.5 \text{ kW}$$

2.7.3.1 Use of PVWatts for Array Sizing

PVWatts, developed by NREL, is an online solar estimate calculator (http://pvwatts.nrel.gov/version_5_3.php) that can be used to find sun hours for a particular location in order to size the solar array. It is a data production tool that calculates how much irradiation falls at the earth's surface at a certain time of the year at a certain location. This data can then be used to work out how much solar energy will be obtained from solar panels installed at a specific location.

The calculation of sun hours using PVWatts at a particular location can be accomplished in three clicks by using the online calculator. The first click of the "GO" button on the calculator's home page is used after entering either the zip code or address of the business or home where one is interested in. This click will take the user to the solar resource data which will be the closest location for TMY2 (Typical Metrological Year second edition), TMY3 or an international file. TMY2 is based on 239 stations collecting data between 1961 and 1990. TMY3 is based on data from 1020 locations in United States, US Virgin Islands, Puerto Rico and Guam which was derived from the 1976–2005 and 1991–2005 periods of record.

The second click of the "Go to system information" arrow will take the user to a few editable fields. For determining the number of sun hours, one has to modify three fields. The first field to modify is the array type which determines the mounting structure of panels being fixed or the one that tracks the sun on one or two axes. The second field to modify is the array's tilt angle from horizontal that the solar panels will be at. The third field is the array azimuth which is the solar array's orientation with respect to true south at 180 degrees.

The third click of the "Results" arrow will calculate the sun hours under the "solar radiation" column in kWh/m^2 per day for each month and annual average. The result also gives AC energy in kWh, energy value for each month and averaged over the year.

2.7.4 Sizing of Charge Controller

Sizing of the charge controller is performed after the design of the PV array in order to safely handle and control the array's incoming power to prevent over-charging the batteries bank. This is achieved by selecting a charge controller based on maximum array watts and nominal battery voltage. Charge controllers are rated and sized by the solar panel array current and system voltage. They are either 12-, 24-, and 48-V controllers with amperage ratings normally from 1 to 60 amps. Traditionally, the nominal voltage of the battery and solar panel array would be the same and one chooses the same voltage for the charge controller. However, charge controllers based on MPPT charging algorithms are used to get the most power out of PV arrays and maintain the highest battery charge. This charge controller allows for a PV array with a much higher voltage than the battery bank's voltage and will automatically and efficiently convert the higher voltage down to the lower voltage so panels, battery bank and PV charge controller can all be equal in voltage. Another advantage of a higher PV array and ability to lower the battery bank voltage is to keep wire size and cost down which can sometimes be 100 feet away from the charge controller. This can also reduce the number of series fuses and the size of the combiner box.

Solar charge controllers are rated and sized by the solar array current which delivers the maximum total PV input current to the controller. This current, also known as maximum source circuit current, depends on whether the PV module is connected in series or parallel. According to Section 690.8 of NEC Code, maximum source circuit current is calculated by multiplying the rated short circuit current I_{sc} by 125%. This multiplier is a safety factor which also takes into account increased irradiance value and the ability of the module to produce more than the rated current.

$$\text{Maximum source circuit current} = \text{Module circuit current} \times \text{modules in parallel} \times \text{safety factor}$$

Maximum source circuit current for I_{sc} of 5.38 A with two modules in parallel is $5.38 \times 2 \times 1.25 = 13.5$ A which should be used to select a charge controller.

2.7.5   Sizing the Inverter

An inverter in a standalone PV system is needed to power AC loads consisting of different appliances. For sizing the inverter the input rating should never be lower than the total power required by the AC loads and must have the same nominal voltage as the battery. While sizing the inverter, it is important to consider a situation where all the appliances and loads are running simultaneously and the inverter should be able to handle the highest continuous power load. The inverter has to have extra capacity to start for these surges which can be two to three times higher than the continuous load. There are also additional loads which need to be accommodated such as surge loads that occur at the start of electric pumps or electric motors for air conditioning, wells, and machine tools. To handle the AC load, the inverter size has to be enlarged by 25–30% of the total AC load. For surge requirement, the capacity of an inverter should be increased to a minimum of three time the capacity of AC load and must be added to the total inverter capacity. In order to size the standalone inverter, a minimum power size is needed which can be calculated by:

$$\text{Inverter minimum power size} = \text{Power of all AC appliances} \times 1.25$$

The factor 1.25 is safety factor for continuous power load.

Power of all AC appliances = Sum of all AC appliances wattages that might run simultaneously as a base load + wattage attributed to appliances that might surge.

In standalone PV systems, inverter input circuits are the conductors between the inverter and the battery bank while inverter output circuits are the AC conductors from the ultimate connection to the AC distribution system.

To determine the conductor size and over-current protection devices (OCPD), Section 690.8(A) (4) of the NEC Code shows the calculation for the highest input current of a standalone inverter. Accordingly, this current which helps determine the conductor size and OCPD rating between the batteries and the inverter is obtained by dividing the inverter's continuous power output rating by its lowest DC operating voltage and then multiplying by the inverter's rated percentage efficiency under those conditions. The percentage efficiency is usually given on the nameplate but can be assumed to be 90% if it is not available. According to section 690.8(A) (3) of the NEC Code, inverter output current is equal to the continuous output current marked on the inverter nameplate.

To determine the minimum OCPD, Imax for a given conductor is multiplied by 1.25 because Section 690.8(B) (1)(a) of the 2011 NEC Code states that OCPD shall be sized to carry not less than 125% of the maximum current.

The resulting continuous current (I_{cont}) is the minimum OCPD required to protect the conductor in the circuit and the minimum rating of all terminals used to make the wiring connections.

$$OCPD = (I_{inverter})(1.25)$$
$$OCPD = (25A)(1.25)$$
$$OCPD = 31.25A$$
$$I_{inverter} = \text{inverter output circuit current}$$

The array-to-inverter ratio defines the relationship between the array's nameplate powers rating at STC to the inverter's rated AC output. As an example, a system with a 120-kWdc array feeding a 100-kWac inverter has an Array-to-Inverter Ratio of 1:2. Until recent years, due to the high cost of modules, PV systems were designed to maximize energy production per PV module. This approach typically resulted in oversizing ratios between 1:10 and 1:25, depending on the project location and design specific DC loss factors such as tilt angle, orientation, mounting method, DC wiring losses, mismatch and soiling. With falling module prices, project financials have changed in favor of higher array-to-inverter ratios. It is important to understand why systems are being oversized, the technical considerations relating to oversizing, and the impact of oversizing on inverter life.

2.8 MODELING AND SIMULATION OF PV SYSTEMS AND SOFTWARE TOOLS

As the photovoltaic industry is undergoing a rapid growth and more PV systems are installed, there is also a greater need to understand the capabilities, limitations, and potential of current and future PV systems. This can be achieved by using modeling and simulation software to design, analyze and estimate not only the PV systems but also the cost of production of energy. These modeling and simulation tools can be classified into prefeasibility study, design and analysis, sizing and dynamic behavior of the system. It is desirable that simulation software should address consideration of simulation and modeling capabilities, hardware and software issues, and consideration of input and output. Many software tools both from private industry and government agencies have been developed and the availability is becoming common as the cost is reduced and easy to learn languages are developed.

In the United States, NREL and Sandia labs have developed software tools on various aspects of PV systems. Some of these free tools are available online which can help in the prefeasibility study of the system. Both NREL and Sandia Lab has published a list of software tools and related material on renewable energy.

NREL has published "Models and Tools" (http://www.nrel.gov/analysis/models_tools.html) which are developed or supported by them to assess, analyze and optimize renewable energy systems. Most of these tools can be applied on a local, regional, global, or project basis. NREL has also developed models and tools specifically for the energy professionals as well as for consumers.

The Sandia National lab has published "Models Used to Assess the Performance of Photovoltaic Systems" (http://prod.sandia.gov/techlib/access-control.cgi/2009/098258.pdf) in the area of PV performance model to estimate the power output of PV systems which include solar panels, charge controllers, inverters, batteries and other components. The list includes models developed and used by researchers at the Sandia National Laboratories as well as models used by other researchers for evaluating PV system's performance.

In the European Union, the University of Applied Science in Munich, Germany has published "PV software Seeker" (http://www.mike-zehner.de/downloads/PV-Softwarekatalog-V1.2-20110210.pdf) which lists 19 pages of PV software in English and German language.

PV resources (http://www.pvresources.com/SiteAnalysis/Software.aspx) has listed over 50 photovoltaic software tools in the following categories.

- Economic evaluation tools
- Photovoltaic industry related tools
- Analysis planning tools
- Smart phone apps
- Monitoring and control tools
- Site management tools
- Site analysis tools
- Solar radiation maps and data
- Online tools

Given below is the description of some of the simulation and modeling software packages used for PV systems.

2.8.1 Hybrid Optimization Model for Electric Renewables

Hybrid Optimization Model for Electric Renewables (HOMER) is a hybrid computer simulation model developed in 1992 by NREL that simplifies the task of evaluating design options for both standalone and grid-connected PV systems for remote and distributed generation applications. It was named after the classical Greek poet Homer. NREL has released more than 42 versions of the program and there are over 100,000 users in 193 countries. This is based on hour to hour simulation and it gives possibilities to control the battery status and to determine the sizing of the batteries. HOMER's optimization simulation allows the user to evaluate the economic and technical feasibility of a large number of technology options and understand tradeoffs and availability of different technologies and energy resources. The software is capable of comparing multiple system configurations as well as battery types using KiBaM (kinetic battery model) code for battery life modeling. This model can simulate and optimize standalone and grid connected or hybrid systems by incorporating any combination of PV arrays, biomass power, fuel cells, hydro power, internal combustion engine generator, micro turbine, wind turbines, as well AC/DC converters, battery, reformer, and hydrogen tank serving both electric and thermal loads. The software is based on 1 year time-period with hour to hour simulation using a minimum time-step of 1 minute which allows it to determine the sizing of batteries and control its status.

HOMER can be downloaded free from the NREL website (www.nrel.gov/homer). NREL also provides classroom and individualized training in the use of HOMER and can customize the software on the request for particular applications or perform specific analyses. NREL's international program continues to provide training and assistance in incorporating renewable energy into rural electrification initiative of developing countries.

2.8.2 Photovoltaic Software

PVsyst is a photovoltaic software developed by the Energy Group at the University of Geneva, Switzerland, in the mid 1990s. It is a widely-used PV software throughout the world with over 460 universities and corporations and more than 1000 users in 55 countries. Many parameters in this software can be modified by users and it is generally considered as the reference tool in the field. PVsyst is a dedicated PC software which integrates prefeasibility, sizing, and simulation support for PV systems and can be used at any location that has solar insolation and meteorological data. The user selects different components from a database after defining the location and loads, and the software calculates the size of the system. The evaluation mode version of PVsyst 6.38 (May 13, 2015) can be downloaded free of cost for 1 month. PVsyst presents results in the form of a full report, specific graph, and tables.

PVsyst is able to import irradiation data from NASA, PVGIS, personal data and meteorological data from different sources in hourly and monthly values.

It also includes a 3-D shading tool which allows the user to draw a structure with PV arrays and observe potential shading impacts from simulated obstructions as well the ability for the user to input known parameters and coefficients, provided that the measured data is available for both PV modules and inverters. Additional information on the availability and maintenance of the software can be obtained from their website: http://www.pvsyst.com/en/

2.8.3 Transient System Simulation Program

Transient system simulation program (TRNSYS) for modeling and simulation of PV systems was developed by the University of Wisconsin, Madison, United States, and has been commercially available since 1975. This simulation software continues to be developed by the international collaboration of the United States, France and Germany. TRNSYS has a modular structure and is based on a system description language in which the user specifies the components that constitute the system and the manner by which they are connected. This simulation software consists of a simulation engine, a graphical interface and a library of components ranging from various building models to standard HVAC equipment to a renewable energy and emerging technologies. It also includes a special library of components for simulation of renewable energy-based power generation systems such as PV systems, wind turbines, batteries, and fuel cells. The software has the ability to interface with other simulation tools, addition of mathematical models and add-on components and has the capability of building multizone models. TRNSYS has been successfully used for detailed analyses of the solar systems whose behavior is time dependent giving calculations being conducted very quickly in a few seconds. TRNSYS 17.1 was released in Jun. 2012 and a demonstration model is available on their website.

2.8.4 Photovoltaic Geographical Information System

Photovoltaic Geographical Information System (PVGIS) is a free online PV energy calculator that was developed by the joint Research Center in Italy and funded by European Commission. This software tool can be used to estimate the expected performance of standalone and grid connected systems and its usage will provide the following:

- Yearly potential electricity generation of the PV system at a certain tilt, orientation, and type of PV module
- Optimum pitch of PV arrays based on geographical location and height above sea level
- Estimated power output of the proposed PV systems
- Amount of solar radiation available at a given location

In case of standalone systems, PVGIS (http://re.jrc.ec.europa.eu/pvgis/apps4/pvest.php) can be used to determine the optimal capacity of a battery

FIGURE 2.12   Map driven user interface of PVGIS (free online PV energy calculator).

which is critical to the battery storage of standalone PV systems. This is useful to get a good assessment of the energy power required to match the electrical needs in remote areas not connected to the grid. The areas covered by PVGIS is Europe, Africa and remote areas in Asia including China, India and Indonesia. It is available in English, French, German, Italian and German languages. PVGIS is easier to use because of its capability of employing Google map applications. To start, the user has to select a location of interest from the map driven user interface as shown in Fig. 2.12.

After determining the location, various specifications such as installed power capacity, panel mounting types, system losses, and others options can be modeled using dropdown menus and tick boxes. Output options for displaying the results such as graphs, text only and pdf format can also be selected by the user. After selecting all the required options for the proposed system, the performance analysis will be run by clicking on the "calculate" button and results are presented within a table and on two optional graphs.

2.8.5   PVWatts Calculator

PVWatts is a web-based calculator which estimates the monthly and annual electricity production as well as the cost of PV systems using an hour-by-hour simulation over a period of 1 year. It was developed by NREL in 1999 and uses typical meteorological year (TMY) data for the selected location at any orientation throughout the world. NREL released a new version of PVWatts 5 in Sep. 2014 which replaced older versions of PVWatts V1 and V2. It was updated on

Jul. 2015 to fix a problem that resulted in overprediction of system energy. It was updated again on April 22, 2016. PVWatts is a useful online tool for conducting preliminary studies of PV systems by determining the solar radiation incident of the PV array and the PV cell temperature for each hour of the year. The DC energy for each hour is calculated from the PV system's DC rating and incident solar radiation. It is then corrected for the PV cell temperature. The AC energy is calculated by multiplying the DC energy with the overall DC-to-AC derate factor and adjusting for inverter efficiency as function of load. Hourly values of AC energy are then summed to calculate monthly and annual AC energy production.

NREL has developed several sophisticated tools including the System Advisor Model (SAM) which allows users to model PV systems in much greater details and more accurate predictions. SAM is a free desktop application tool that includes detailed economic analysis for commercial, residential, and utility-based systems and performance models for solar water heating, geothermal systems, and concentrating solar power.

For PVWatts usage, users can select a location and choose to use default values or their own system parameters for size, electric cost, array type, tilt angle, and azimuth angle. PVWatts will use the address of the required system's location or zip code, or geographical coordinates on its website (http://pvwatts. nrel.gov/version_5_3.php) to identify solar resource data available at or near the system's location. By putting any one of these three information in the United States will give result from default value already used. This location may be anywhere in the United States, or in some other part of the world. To estimate the monthly and annual electricity production of a photovoltaic system the following physical parameters require values for the input. However the default systems parameters provided in the system can also be used to conduct a preliminary study if default values are acceptable, such as:

- System DC size or DC rating
- DC to AC derate factor
- Type of PV array whether fixed or tracking
- PV array tilt angle
- System losses
- PV array azimuth angle
- Local electric costs

BIBLIOGRAPHY

[1] Affordable Solar's Off-Grid Load Estimator. <http://www.affordable-solar.com/Learning-Center/Solar-Tools/Off-Grid-Load-Estimator> <http://www.nabcep.org/wp-content/uploads/2012/08/NABCEP-PV-Installer-Resource-Guide->.

[2] D.F. Al Riza, S.I. Gilani, M.S. Aris, Measurement and simulation of standalone system for residential lighting in Malaysia, J. Hydrocarbons Mines Environ. Res. Volume 2 (Issue 1) (2011). June 2011, xx–xx <http://jhmer.univ-rennes1.fr>.

[3]   Andrew, Solar PV & Remote, Distributed Microgrids Poised to Improve Living Conditions for Millions. CleanTechnica. <http://cleantechnica.com/2012/01/04/solar-pv-remote-distributed-microgrids-poised-to-improve-living-conditions-for-millions/>, 2012.

[4]   J. Ayre, New CIGS Solar Cell Record—21.7% CIGS Cell Conversion Efficiency Achieved At ZSW. <http://cleantechnica.com/2014/09/27/new-cigs-solar-cell-record-21-7-cigs-cell-conversion-efficiency-achieved-zsw/>, 2014.

[5]   J. Balfour, M. Shaw, S. Jarosek, Introduction to Photovoltaic: The Art and Science of Photovoltaic, Jones & Bartlett Learning LLC, Burlington, MA, 2013.

[6]   M. Bingeman, Nuvation, B. Jeppesen, Improving Battery Management System Performance and Cost with Altera FPGAs. Industrial, Automotive and Broadcast System Solution Engineering, Altera Europe. Altera Corporation. <https://www.altera.com/content/dam/altera-www/global/en_US/pdfs/literature/wp/wp-01247-improving-battery-management-system-performance-and-cost.pdf>, 2015.

[7]   B. Brooks, Inspecting Photovoltaic (PV) Systems For Code-Compliance. Brooks Engineering. <http://www.pge.com/includes/docs/pdfs/shared/solar/solareducation/inspecting_pv_systems_for_code_compliance.pdf>.

[8]   I. Clover, First solar raises bar for CdTe with 21.5% efficiency record. Global PV markets, industry & suppliers, markets & trends. Res. Dev. <http://www.pv-magazine.com/news/details/beitrag/first-solar-raises-bar-for-cdte-with-215efficiency-record_100018069/#ixzz3j5UWDXw1>, 2015.

[9]   Cadmium Telluride, Energy.Gov, Office of Energy Efficiency & Renewable Energy <http://energy.gov/eere/sunshot/cadmium-telluride>.

[10] Copper Indium Gallium Diselenide, Energy.Gov, Office of Energy Efficiency & Renewable Energy. <http://energy.gov/eere/sunshot/copper-indium-gallium-diselenide>.

[11] J. Dunlop, Photovoltaic Systems, second ed., American Technical Publishers, Orlando Park, IL, 2010.

[12] J.A. Fecteau, Inverters in Photovoltaic Systems, Electrical Connection: A Supplement of the Code Authority. <http://www.ul.com/global/documents/corporate/aboutul/publications/newsletters/electricalconnections/January12%20.pdf>, 2012.

[13] L. Folsom, Module Model: The Semiconductor Physics that Lies at the Heart of Performance Modeling. <https://www.folsomlabs.com/modeling/module/module_model>, 2014.

[14] C. Freitas, High–Capacity Battery Banks: Product Selection and System Design Fundamentals. <http://solarprofessional.com/articles/products-equipment/batteries/high-capacity-battery-banks>, 2012.

[15] W.A. Hermann, Quantifying global energy resources, Energy 31 (12) (2006) 1685–1702 <http://www.sciencedirect.com/science/article/pii/S0360544205001805>.

[16] M. Hartmann, M. Zehner, O. Mayer, PV Software Seeker. Version 1.2—Status: 20110210. Arbeitsgruppe im Studiengang. <http://www.mike-zehner.de/downloads/PV-Softwarekatalog-V1.2-20110210.pdf>, 2011.

[17] Handbook of Secondary Storage Batteries and Charge Regulators in Photovoltaic Systems. <http://www.azsolarcenter.org/tech-science/technical-papers/battery-handbook-for-pv-systems.html> <http://www.azsolarcenter.org/images/docs/tech-science/papers/batteries/ch4.pdf>, 2015.

[18] HOMER (Hybrid Optimization Model for Electric Renewables). The Micropower Optimization Model. <http://www.nrel.gov/docs/fy04osti/35406.pdf>.

[19] C. Katz, Will New Technologies Give Critical Boost to Solar Power? Environment 360, Reporting, Analysis, Opinion and Debate. <http://e360.yale.edu/feature/will_new_technologies_give_critical_boost_to_solar_power/2832/>, 2014.

[20] G.T. Klise, J.S. Stein, Models Used to Assess the Performance of Photovoltaic Systems, Prepared by Sandia National Laboratories, Albuquerque, NM, 2009. <http://prod.sandia.gov/ techlib/access-control.cgi/2009/098258.pdf>.

[21] Lithium–ion safety concerns, Battery university. <http://batteryuniversity.com/learn/article/ lithium_ion_safety_concerns>.

[22] R. Meck, Off-Grid System Sizing. Affordablesolar <http://www.affordable-solar.com/ Learning-Center/Solar-Basics/Off-Grid-System-Sizing>   <http://data.energizer.com/PDFs/ nickelmetalhydride_appman.pdf>.

[23] R. Mayfield, PV Circuit Sizing & Current Calculation. Home Power <http://www.home-power.com/articles/solar-electricity/design-installation/pv-circuit-sizing-current-calcula-tions>, 2014.

[24] P. Manimekalai, S. Raghavan, An overview of batteries for photovoltaic (PV) systems, Int. J. Comput. Appl. (0975–8887) Volume 82 (No 12) (2013) 28–32.

[25] Nickel-Metal Hydride, Handbook and Application Manual. Energizer Battery Manufacturing Inc. I 800-383-7323 (USA-CAN).

[26] NREL Solar Advisor Model. <https://sam.nrel.gov/>.

[27] NREL: Dynamic Maps, GIS Data, and Analysis Tools—Solar. <www.nrel.gov/gis/solar. html>, 2015.

[28] NREL's PVWatts® Calculator. <http://pvwatts.nrel.gov/>.

[29] A. Pradhan, S.M. Ali, S.P. Mishra, S. Mishra, Design of Solar Charge Controller by the use of MPPT Tracking system. www.ijareeie.com <http://www.ijareeie.com/upload/october/6_ Design%20of%20Solar%20Charge%20Controller.pdf>.

[30] Planning & Installing Photovoltaic Systems, A Guide for Installers, Architects and Engineer. Published by Earthscan ISBN 1844074420, 9781844074426.The German Energy Society, 2008.

[31] Photovoltaic and Distributed Generation. <http://envirostewards.rutgers.edu/Lecture%20 Resource%20Pages/Energy%20resources/PhotoVoltaics/Photovoltaics%20-%20PV%20 Basics.htm>.

[32] PVSYST 4.33. <http://www.pvsyst.com/en/>.

[33] PV Resources, Photovoltaic software and Tools. <www.pvresources.com>, 2015.

[34] Photovoltaic Geographical Information System (PVGIS). Geographical Assessment of Solar Resource and Performance of Photovoltaic Technology. Joint Research Center. <http://re.jrc. ec.europa.eu/pvgis/>.

[35] S. Qazi, F. Qazi, Nanotechnology for photovoltaic energy: challenges and potentials, in Handbook of Research on Solar Energy Systems and Technologies, IGI Global, Engineering Science Reference, Hershey, 2013.

[36] Remote Microgrids, Village, Island, Industrial Mine, and Mobile Military Power Grids. Navigant Research   Webinars.   <https://www.navigantresearch.com/webinar/remote-microgrids-2>, 2012.

[37] J. Sanchez, PV Array Sizing for kWh. Home Power, Issue#152. <http://www.homepower. com/articles/solar-electricity/design-installation/pv-array-sizing-kwh>, 2013.

[38] K. Schneider, New World Record For Solar Cell Efficiency at 46% Frech-German Cooperation Confirm Competitive Advantage of European Photovoltaic Industry. Bernin, France and Freiburg, Germany. <http://www.ise.fraunhofer.de/en/press-and-media/pdfs-zu-presseinfos-englisch/2014/press-release-new-world-record-for-solar-cell-efficiency-at-46-percent.pdf>, 2014.

[39] Solar America Board for Codes and Standards, Underwrite Laboratories. <http://www.solar-abcs.org/codes-standards/UL/index.html>.

[40] Solar photovoltaic software, Michigan Tech's Open Sustainability Technology Lab. <http://www.appropedia.org/Solar_photovoltaic_software>, 2010.

[41] Solar Energy. <http://www.inforse.org/europe/dieret/Solar/solar.html>.

[42] TRNYS (Transient System Simulation Program). <http://sel.me.wisc.edu/trnsys/>.

[43] F. Vignola, F. Mavromatakis, J. Krumsick, Performance of PV Inverters. <http://solardat.uoregon.edu/download/Papers/PerformanceofPVInverters.pdf>.

[44] Z. Wei, A smart battery management system for large format lithium ion cells. The University of Toledo Digital Repository: Theses and Dissertations. Paper 769. <http://utdr.utoledo.edu/cgi/viewcontent.cgi?article=1790&context=theses-dissertations>, 2011.

[45] C. Woodford, Lithium-Ion Batteries Explain That Stuff. <http://www.explainthatstuff.com/how-lithium-ion-batteries-work.html>, 2015.

Chapter 3

Mobile Photovoltaic Systems for Disaster Relief and Remote Areas

3.1 INTRODUCTION

Mobile photovoltaic (PV) systems are normally mounted on trailers, personal automobiles, placed on carts, mounted as packs on horses or camels, or shipped in standard shipping containers, and then moved to wherever needed. These are autonomous systems which can be used as a source of power for homes, businesses, hospitals, and schools in the aftermath of disasters and in remote areas. Mobile PV systems consist of solar panels to produce electricity and a charge controller to regulate the voltage and current. These systems also contain rechargeable batteries to store electricity, inverters to change DC into AC, and wiring to connect various components. The components used in mobile PV systems are similar to standalone systems but are configured for the harsh demands placed on mobile systems. Mobile PV systems offer an advantage in situations when conventional fuel would be difficult to deliver, costly, or causes pollution. The mobility of photovoltaic systems is an important feature when the supply of electrical power must be moved from one location to another in disaster relief and/or remote areas. These systems are commercially available in different sizes and shapes and can also be implemented as DIY using off-the-shelf components.

Mobile PV systems mounted on trailers are the most commonly used systems for bringing services to people in the aftermath of a disaster and in remote areas. The advantages of these systems include powering many applications such as lighting systems, pumps for clean water, traffic lights, radio and TV stations, emergency clinics, power tools, and telecommunication equipment. Mobile PV systems are also used for outdoor events, construction sites, and reliable power for farms is either not available or is very expensive due to being designated as a remote area. These systems were first used in the disaster relief of Hurricane Hugo in 1988 to provide power for various usages. These systems on trailers have been used in many disasters including Northridge earthquake (1994), Hurricane Andrew (1992), Hurricane Bonnie (Aug. 1988), Hurricane

Standalone Photovoltaic (PV) Systems for Disaster Relief and Remote Areas.
DOI: http://dx.doi.org/10.1016/B978-0-12-803022-6.00003-4

Georges (Sep. 1988), Hurricane Charlie (2004) and Hurricane Katrina (2005), Haiti (2010) and Superstorm Sandy (2012). Nomadic communities of Kenya's Laikipia and Samburu districts have been using camels for decades to deliver medicine and vaccine in the difficult terrain that lacked roads. However, in 2005, their community's Trust, with the help of two US educational institutions, created a lightweight solar-powered refrigerator that can be strapped to a camel's back. Mobile solar systems have also been used in remote areas to provide electricity and clean water.

3.2 SOLAR-POWERED MOBILE TRAILERS

Solar-powered trailers are often used for providing power to small communities, homes, and businesses in a remote area or an area affected by a disaster in applications such as communications, lighting, water purification, clinics, and shelters. These systems are commercially available in different sizes and can also be implemented using off-the-shelf components. Solar-powered trailers are also used at camping sites, construction sites, and public events that are not connected to the electrical grid. The applications of mobile PV systems in the aftermath of disaster depends on the degree and scale of natural disasters. After Hurricane Hugo (1989), mobile PV generator systems were used to power a community center for 6 weeks. During the Northridge earthquake (1994), PV systems were used to keep some communication links operating and to power homes of southern California residents. During Hurricane Andrew (1992), PV systems were used at medical clinics and shelters and to power street lights and communication systems. The PV systems for disasters in Haiti (2010) were used to provide low-cost, reliable electricity to a poor rural area for schools, community centers, small businesses, and homes to power LED lightbulbs, run a radio or power a small power tool and charge mobile phones. After Superstorm Sandy (2012) in New York, mobile solar generators donated by SolaRover were used to support a Greenpeace clinic, soup kitchen facility, and a gymnasium.

3.2.1 Features of Solar-Powered Mobile Trailers

Although solar-powered mobile trailers are similar to PV standalone systems, these systems have some unique requirements, different than standalone systems. The important design features to consider for trailer are size, load capability, stability and strength of the trailer, and should be based on the following features:

- Compliance with the relevant Department of Transport Standards for use on highways and trailer industry guidelines and practices
- Provide safety and ease of PV panels implementation
- Firm support of PV arrays by locking and making secure
- Roof of the enclosed trailer should have vents to withstand the internal heat
- The folded solar array should be able to withstand winds of up to 60 mph and vibrations during transportation

- The height of the trailer should be enough for a person to stand inside the trailer
- Size of the trailer determines the weight that can accommodate the batteries and axle
- Easy to fold and unfold the solar panels
- Capable of withstanding exposure to severe weather without damage
- Should withstand hazardous conditions
- Smallest footprint for power production

3.2.2   Types of Solar-Powered Mobile Trailers

Solar panels can be used on all three types of trailers:

- Conventional travel trailers
- Fifth wheel trailers
- Basic cargo trailers.

The design of these trailers can be enclosed (shown in Fig.3.1A) to protect supporting electronic components or have an open design (shown in Fig. 3.1B).

A conventional travel trailer is a nonmotorized vehicle designed to be towed by a powered vehicle such as SUV, a pickup truck, or, for smaller units, even by a car. These units offer all the amenities depending on the model and floorplan and up to 10 people can sleep in it. In some countries, the campers are restricted to be parked at designated sites where fees are payable. A utility trailer is attached to a motorized vehicle by a hitch which is usually attached to the chassis of a vehicle for towing. The commonly used hitch for the trailer is a 2-inch ball hitch. The utility trailer has wheels that can be constructed as an enclosed trailer or open air flatbed trailer. A utility trailer can be built with a special built-in equipment or shelves to carry specialty equipment like solar panels or used for recreational purposes. The normal length of a travel trailer measures between 14 and 36 feet and the average length varies between 21 and 28 feet. The normal width range is from 6 to 8 feet (Not including slide outs).

(A)   (B)

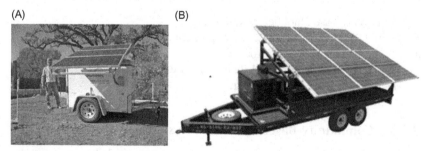

FIGURE 3.1   (A) Enclosed trailer; (B) Open trailer.

A fifth wheel trailer is designed to be affixed and towed by a pickup truck with a special fifth wheel hitch installed in the bed of the pickup. In the travel trailer, the hitch is located on the rear bumper. The size of the fifth wheel trailer ranges from 20 to 40 feet and are available in a variety of floorplans. The floorplans include new Bunk Bed Models, and the use of fifth wheel hitch differs from ball/socket type commonly used in smaller trailers. Fifth wheel trailers can provide the largest living area with all the comforts of home.

A basic cargo trailer is a flatbed trailer often used to transport goods and livestock for short distances. It is normally attached to another motorized vehicle that pulls it along the road or in the field. The long-haul trailers are used for bringing cargo over a longer distance and comes with a maximum length of 17 m with many tires for extra support. In the United States, all roadworthy trailers must have functioning brake lights and appropriate license plates.

It is recommended that with the use of cargo/vehicle, the height of the trailer should be at least 6 feet when enclosed. Plywood is a good material to use on the inside of the trailer and rib spacing makes it easy to attach equipment. Ventilation should be used to reduce heat buildup. Strong and reliable latches and doors such as a ramp and single hung type should be used to overcome wind and vibrations on the highway.

3.3 EXAMPLES OF MOBILE PV SYSTEMS FOR DISASTER RELIEF AND REMOTE AREAS

In the aftermath of disasters, electricity is needed to immediately provide emergency services and reconstruction of homes and businesses. This need starts from communications for emergency personnel to request assistance, supplies, and information. Homes and businesses require electricity, water, sewerage, and deployment of equipment to restore essential services. Shelters and medical clinics are needed to temporarily accommodate and treat the affected people. Solar-powered mobile systems are able to provide these services as refueling might be difficult or impossible due to damaged roads or ports. In remote areas that have no access to grid, all these services are needed on a permanent basis or for a longer period. Listed below are some of the examples of providing electricity to various needs in the aftermath of disasters and remote areas:

- Community-based mobile PV systems
- Mobile PV system for power, water purification, and communications
- Mobile PV medical clinics for remote areas
- Mobile PV systems for small shelters and illumination
- Hand-driven solar generation cart

3.3.1 Community-Based Mobile PV Systems

A community is often faced with a loss of electricity to run daily services in the aftermath of a natural disaster. Mobile PV systems can play an important role in the recovery of natural disasters and help in emergency operations. One example of a community-based mobile solar system using a trailer is the SunBlazer which was used in the aftermath of Haiti earthquake in 2010. It was funded and delivered by a team of *Institute of Electrical and Electronics Engineers* (IEEE) volunteers to provide low-cost, reliable electricity to six rural areas after the devastating earthquake in Haiti. The program was implemented by creating a nongovernmental organization (NGO) Sirona Cares Foundation (a community solutions initiative (CSI) by IEEE) who helped to operate, manage and develop customer base in partnership with NGO Sirona–Haiti. A new Haitian not-for-profit entity, Sirona-Haiti, was formed to start in-county businesses including manufacturing. The SunBlazer consist of six solar panels of 245 W each arranged in two parallel banks of three as shown in Fig. 3.2.

The station generates 1.5 kW and is built from a trailer chassis of 8 feet in length, 5 feet in width, and 1.5 feet in height. It carries 40 portable customer home battery kits, each consisting of 12 V, 18 Ah portable battery packs. The charging station also contains four large lead–acid house storage batteries, a charge controller, auxiliary rectifier, and 20 subchargers for 40 home customers serving an estimated 1400 people. A special maximum power point tracking charge controller is connected between the PV panels and the batteries for smooth charging the house batteries. Additional trailers are used for large base-load customers. Four large batteries in the trailer are charged from the solar panels which can then be used to charge smaller home packs. The charged home packs are then carried to homes where they are needed for powering cell phones, LED lightbulbs, radios or small power tools. Once the home battery pack is 50% depleted, it automatically cuts off to preserve a long battery life. The depleted battery must then be carried back to the trailer station where it is replaced by another battery or the homeowner's battery receives a recharge. Appliances such as small refrigerators, water purification systems or pumping

FIGURE 3.2   SunBlazer trailer. *Photo Courtesy Ray Larsen, IEEE Smart Village.*

stations can also be powered by several battery packs. Surplus power from the charging station is sold or converted to be used by the local community centers, small businesses, health clinics, or schools.

The IEEE Community Solution Initiative group in Feb.–Mar. 2012 delivered nine more SunBlazer trailers to Grand Goave, a small city of nearly half a million inhabitants in Haiti, most of which was destroyed by the 2010 earthquake. These stations are equipped with 80 home kits per trailer to serve 4300 people. The first six SunBlazer trailers for Haiti were built in United States while the second nine were partially constructed with the help of Sirona–Haiti. The Sirona–Haiti plans to manufacture 80–100 units per month in Port au Prince with the help of new funding by USAID with the goal of serving about 1 million people in the next 5 years. With the addition of these nine SunBlazer trailers, Haiti is being served successfully with affordable electricity by the 15 charging stations that use mobile trailers. In addition, it has benefitted economically with jobs, investment in local economy, and knowhow of starting small businesses. SunBlazer trailers have also given a new impetus to the microgrid solution to the power generation which might be more appropriate to the remote areas having no access to the electric grid. Microgrid solutions are also more secure and easy to handle at the time of disasters and in remote areas.

Encouraged by the success of SunBlazers, the creators of CSI/IEEE have already started to work with energy-poor areas in Africa and south Asia and plan to provide electricity to 40 million people by 2020. To accomplish this, CSI plans to support ten new startup businesses each year using seed money raised through an ongoing special fund under the IEEE foundation. The startups will raise $10 million through grants, loans or venture funding to produce enough trailers to reach over a million people in 5 years. The potential market for supplying these trailers continue to grow globally where more than 1.4 billion people do not have access to electricity and 2.4 billion people lack access to fuels for cooking and heating. It is a huge challenge to commence a new startup in developing countries where problems of starting business are compounded with low income and national bureaucracy.

3.3.2   Mobile PV Systems for Power, Water Purification, and Communications

Provision of power, clean water, and communications is essential in the emergency response of natural disasters and remote areas with no access to grid. Many manufacturers have developed PV based systems to provide these services.

One such system to provide electrical power, communications and clean water to remote areas, is mobile system GSW-7000 developed by Providence Inc. This system (shown in Fig. 3.3) can be powered by both solar or a windmill which can also be used in the aftermath of disasters. The systems consist of three units: a power unit, communication unit, and water filtration system, all in a single platform. GSW-7000 consist of 16 solar panels, a wind turbine,

FIGURE 3.3   Water purification, power and communication system powered by solar and wind energy. Providence Inc. <http://providencetrade.com/mobile-solar-and-wind-powered-energy-water-and-communications-system>.

a 106-foot telescopic communication tower, an inverter, rechargeable batteries, and an optional solar-powered water purification system. The entire system is integrated into a trailer that measures 34 feet (10.4 m) in length, 7 feet 3 inches (2.2 m) in width and 7 feet 3 inches (2.2 m) in height that can be towed by a standard duty pickup truck. The system is equipped with an onboard hydraulic mechanism which can automatically deploy the solar panel's communication tower in less than 30 minutes. The solar panel arms and the tower can be packed for storage and transportation. The details of each of the units are given below:

> The Power unit is used to generate electricity by using both solar panels (8 panels for 2300 W and 16 panels for 4600 W) and wind turbine which produces 2.4 kW in a 29 miles/hour speed of wind. During the time of no sun or wind 24 rechargeable batteries rated at a power rating of 1500 Ah at 48 VDC (72 kWh) are used for storage. These batteries are based on Absorbent Glass Matt (AGM) technology which is deep cycle, sealed, lightweight, spill proof, maintenance free, and capable of delivering high currents on demand. For increasing the life of the batteries, an onboard power management optimal charge and discharge is used. An inverter of 6 kW rating is used to convert DC output of solar panels into 240/120 V AC both single phase and multiphase. Auxiliary power generator option for sustainable power output from 3 to 12 kW is available with natural gas, LP gas, gasoline or diesel. NASA has used GSW-700 to recharge the batteries for several of their robotic vehicles in Desert RATS (Desert Research and Technology Studies) program which simulate and evaluate technology, human robotic systems, and extravehicular equipment for future human exploration in space.
>
> The Communication unit of the solar-powered mobile system consists of an expandable tower which provides a flexible platform for communication, surveillance, and lighting. The tower contains a range of mounting points for a variety of communication systems including cellular, microwave, satellite, WiMAX, WIFI and can be extended to 106 feet (32 m). Eight

outriggers deployed hydraulically provide stability to trailer and tower in winds in excess of 110 miles/hour (177 km/hour). High grade, drawn over mandrel (DOM) carbon steel is used to construct the tower for allowing maximum payload of 660 lbs (300 kg). The entire system requires minimal maintenance and can be remotely controlled and monitored from a single control panel. The tower hydraulic system hitch is pintle ring.

The optional solar-powered water purification system provides clean drinking water from contaminated water found in rivers, lakes, and ponds in remote areas. The unit also has a reverse-osmosis purification system for making potable water from both brackish (semi-salty) and sea water. The water purification system can produce about 6336 gallons of water during 8 hours of sun alone. However with batteries charged with solar panels or wind turbine, the potential output for 24 hours increases to 19,000 gallons a day.

Other manufacturers like NuSpectra have similarly developed solar power trailers for remote power, communications, lighting, and monitoring which can be employed in the aftermath of disasters and remote areas. The system as shown in Fig. 3.4 consists of 4.3 kW solar panels, 30 feet telescopic mast, eight 150 Ah sealed lead–acid batteries, and 1.5 K volt, 120 volt AC inverters.

This system provides 24 V DC output and converts it into AC using 1.5 K volt, 120 volt AC inverters for 30 hour working time. The system uses four adjustable long life 100 W LED lights for lighting. For remote monitoring, two PTZ (pan tilt zoom) megapixel cameras with infrared are used for nighttime and two fixed 2.0

FIGURE 3.4 Mobile PV system for power, communication, and lighting.

megapixel cameras are used for day time. It achieves remote IP communications using cellular, satellite, wireless ISP, and wired solutions. Optional equipment includes cellular radio, wireless internet satellite provider, satellite internet, digital road signs, construction security camera package, traffic monitoring equipment, point to point radio, wide range of IP cameras and security, backup diesel power and wind turbine. The measures 8.6 feet in length, 8.2 feet in width and 6.4 feet in height and weighs 2600 lbs all made of steel and aluminum.

The system can be used for illuminating disaster–affected areas, CCTV surveillance, remote webcam platform, and remote communication. It is clean, noise free, provides continuous night operation and can handle multiple cloudy days. It can be used without any attendance, needs very little maintenance and its power can be managed remotely.

3.3.3   Mobile PV Systems for Medical Clinics and Remote Areas

Many remote areas in developing countries do not have access to medical facilities and the use of reliable electrical power. According to the World Bank, more than 60% of people in Sub-Saharan Africa live in rural areas without access to medical facilities or lack resources to reach clinics for proper medical care. In South Africa, approximately 20% of the population can afford to make use of private medical services while the remaining 80% of the population are served by the struggling public health sector. According to a recent report by the World Health Organization, nearly 25% of the medical facilities in eleven Sub-Saharan countries covering 4000 clinics and hospitals either have limited, or no access to, electrical power. Without electrical power, health facilities cannot use many of the basic life-saving devices or equipment such as vaccine refrigerators, pumps or heat waters, turn on the lights to deliver babies, charge a cell phone or perform surgeries. In view of these medical issues, Samsung Inc. in 2013, launched its first solar-powered mobile health center for Africa in Cape Town, South Africa.

The mobile health center is a 7-meter truck which uses PV panels and provides power for lighting, small appliances, and a TV monitor that displays public information to the patients waiting to be treated. The unit also uses inbuilt generators to power the other equipment used in the truck. The health center is fully equipped to perform surgery on ears and eyes. Emphasis is placed on screening people to establish conditions such as high blood pressure, cataracts, diabetes and tooth decay. Samsung has introduced the following solar-powered facilities to make it a digital village:

- Solar-powered telemedical center: A solar-powered telemedical center to gain access to experienced medical specialists in large renowned centers. This center helps to reduce the shortages of qualified doctors in these remote areas and also save patients from traveling to big centers. Its advantages also include quick diagnostics and consultation.
- Solar-powered generator: A solar-powered generator to provide affordable electricity for schools and community centers to power classroom

equipment including notebooks, eboard, printers, lighting, and other needed equipment. The money saved by using affordable electricity can be used in other projects.

- Solar-powered internet school: It is built in a 12-m-long mobile shipping container which is powered by solar panels to power the class room's equipment including an interactive eboard for 24 people. The classroom can also be used as an adult education and community center in the afternoon and on the weekend.
- Solar-powered administration center: To be used as an administration office by staff members

3.3.3.1 Solar-Powered Clinic in A Can

A solar-powered clinic named "Clinic In A Can" was developed in 2012 for remote areas. It can also be used for medical relief in the aftermath of a disaster or the area with no electricity. "Clinic in a Can" is a division of Hospitals of Hope, which is a medical not-for-profit organization founded in 1998 in Wichita, Kansas. The mobile medical clinic consist of a 40-foot shipping container converted into portable, self-sustaining, fully functional medical clinic. The 40-foot standard shipping container measures 8 feet in width, 8 feet and 6 inches in height and weighs 67,200 lbs (30,480 kg). The clinic contains two examination rooms, a laboratory, and a mechanical room, all powered by solar energy. The solar panels mounted on the shipping container, as shown in Fig. 3.5, convert direct sunlight into electricity for powering a three room clinic consisting of medical equipment, lighting, and air conditioning.

The clinic also includes rechargeable batteries which are used to provide power to the clinic for 3 days when there is no sun. The medical clinic has

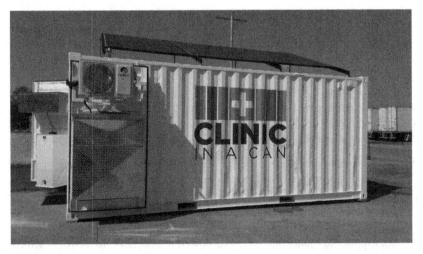

FIGURE 3.5 Solar-powered mobile medical clinic.

telemedicine capability to communicate with experts and consultant for assistance and diagnosis. It is equipped with split system climate control units, a filtered water system, and energy efficient LED lighting.

A Clinic In a Can is transported by trucks and can be shipped anywhere in the world. It is designed to provide safe environment for medical treatment and can survive hurricanes, earthquakes, and other natural disasters. The idea of "Clinic in a CAN," was conceived in 2002, and the first medical clinic was sent to Jefferson parish in Louisiana in 2005, after Hurricane Katrina. In the 2010 earthquake, Haiti received six; and in South Sudan three of these medical clinics were used during the civil war. In 2013, two more medical clinics were sent to Philippines after the typhoon floods. Clinic in A Can has been used in many countries including Haiti, Nigeria, Philippine, Sierra Leone, and South Sudan. The clinic has also been used in United States for disaster relief and isolation wards.

3.3.3.2   Solar-Powered Camel Clinics

In the remote areas of northern Nigeria, nomadic communities have been using camels to transport medicines and vaccine to places with no roads. This area is inhabited by a population of around 500,000 people with a low literacy rate and poor health care. In order to provide some aspect of health care, a Kenyan-based Nomadic Communities Trust (NCT) has been using camels to transport medical supplies in this difficult terrain. These medical supplies were carried in wooden boxes which were then tied to camels with abrasive sisal rope. This method of transportation, although cost effective, had no way of refrigerating vaccines and medicine. In 2005, the NCT partnered with California Arts College of Design's designmatters Department and the Princeton Institute for Science and Technology Materials to create a solar-powered lightweight refrigerator for refrigeration of vaccines and medicines. The solar-powered mini refrigerator as shown in Fig. 3.6A is housed on a bamboo saddle which is lightweight enough to be carried at the back of camels (as shown in Fig. 3.6B) through rough terrain for many miles.

The mini refrigerator is housed on a bamboo saddle and is powered by the mono-crystalline solar panels placed on top of the fridge. The fridge is mounted

(A)   (B)

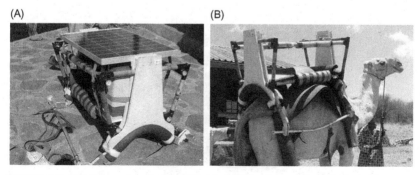

FIGURE 3.6   (A) Solar-powered mini refrigerator; (B) bamboo saddle at the back of a camel.

on top of a lightweight frame of hardened aluminum which can support approximately 300 lbs (136 kg) of cargo. The solar panel with the rechargeable battery also provides power at night and for video equipment sessions' on health education. The mobile clinic was beta tested by using camels from the Bronx Zoo in New York, United States, for a few thousand US dollars. These solar-powered fridges have been tested on camels' backs in Kenya and Ethiopia and can be used by remote rural communities with little roads and poor access to health clinics. The system could be used by any rural communities with access to camels.

3.3.4 Mobile PV Systems for Small Shelters and Illumination

Electrical power, shelters, lighting, and storage space for supplies and tools is often needed in the aftermath of a disaster and remote areas. Mobile Solar Inc. has developed five models of mobile solar generators starting from 4 × 6 foot trailer platform (3 panels, 4.5 kWh/day, 120 V), 5 × 9 foot platform (6 panels, 9 kWh/day, 120 V VAC), 5 × 9 foot platform (6 panels, 9 kWh/day, 120 V/240 VAC) 5 × 16 foot (9 panels, 13.5 kWh/day, 120/240 AC), 6 × 16 foot (12 panels, 18 kWh/day, 120/240 VAC) and the largest system of 6 × 20 foot (16 panels, 24 kWh/day, 120/240 AC).

The largest system, MS-375, shown in Fig. 3.7, is built on 6 × 20 feet trailer and has a large cargo space which can be used as a solar-powered shelter, an office, or a storage for critical supplies in the aftermath of disasters and remote areas. The smaller systems can equally be used to house people, food, emergency supplies, fragile electronic equipment or tools. The mobility of these mobile generators allows serving large number of people at different locations.

MS-375 consists of 16 monocrystalline Suniva PV panels of 270 W each, 37 kWh battery, a MPTT controller and an inverter. The trailer generates 24 kWh/day which can provide power for many community based activities after disasters and in remote areas. The 3500 lbs battery built on 20 foot, 2 axle trailer with gross vehicle weight rating (GVWR) of 10,000 lbs can be towed behind a ¾-ton truck. An inverter (OutBack) rated at 3500 W continuous output and 6000 W

FIGURE 3.7 Solar-powered trailer for small shelter, office or storage.

surge with efficiency of 92–94% is used to convert DC output to distortion-free sinewave AC. A maximum Power Point Tracking system (OutBack FM60/80-amp) is used to ensure that solar panels are operating at its peak point regardless of shading, clouds or age of the system. This system can be used to power loads of 7000 W or greater on a permanent basis for providing telecommunication, water pumping power system. It also provides power to an event and off-grid homes. The system was used to power an event which was organized by Arizona Public Service, where former president Bill Clinton spoke at the Clinton Global Initiative (CGI) Conference in Phoenix in 2013. CGI hosts annual meetings where students, youth organizations, experts in different fields and celebrities meet to discuss and develop innovative solutions to pressing global challenges.

Mobile solar Inc., has also developed three models for solar-powered light towers mounted on wheels that can be used for illuminating disaster affected areas or remote places with no access to grid electricity. Each of these systems, as shown in Fig. 3.8, are equipped with four 50 W high lumen LED flood lights (24,600 lumen output), a 21 foot crank up tower, and is supported for wind speed up to 85 mile/hour by four outriggers.

The tower is installed on a single axle powder-coated steel weighing 3500 lbs, which can be easily folded and towed. One of the models, LT-2435, is also equipped with an onboard inverter/charger that can produce 120 V pure sine wave AC power while allowing the user to charge the 24 V DC battery bank with a generator to the grid. It uses two monocrystalline PV cells, a 3500 W inverter, 120 V AC output (6 kW surge), 30 A MPTT charge controller with digital display, and 4 × 8d AGM 460 AH batteries.

The solar-powered light tower is programmed for pre-dawn and post dusk periods by selecting the number of hours the lights turn on before sunrise and a number of hours lights turn on after sundown with an off period in the middle to save power. It can also be used for all night lighting without any sound. The lighting controller is also programmed to automatically calibrate to changing daylight hours.

(A)   (B)

FIGURE 3.8   (A) Solar-powered light tower; (B) folded solar-powered light tower.

3.3.5 Hand-Driven Solar Generator Cart

A hand driven solar-powered cart is an example of mobile PV system for when a motorized trailer is not available to power small appliances in the aftermath of disasters and remote areas. It can be easily assembled as a DIY project at an affordable cost by anyone with a basic knowledge of electricity. One such system, the SolGen 160, was built by Bill Brooks in 2011 who assembled 4 foot (length) × 4 foot (width) × 4.5 foot (height) cart on the wheels using a 2 × 3 lumber for framing and T-1 siding for the enclosure. Two L-shaped pieces for each panel are built to hold the panel at a 45-degree angle.

A hand-driven solar generator consists of two 80 W solar panels, 210 Ah deep cycle marine battery, 30 A charge controller and an 1100 W power inverter. The battery and controller are enclosed inside the cart which can be moved or repositioned as needed. This system is capable of providing 460 A of power each week to charge a 12 V battery. The DC/AC output from the cart can power a microwave oven, refrigerator, and LED lights in a house, or to power a number of small appliances such as microwave oven, laptop computer, TV, LED lights and small power tools in the aftermath of a disaster and remote areas. To access the battery and other components inside the cart, doors are added at the back and the cart is caulked to prevent leaks. The cart was built using off-the-shelf components.

3.4 CASE STUDY FOR SOLAR CARS

To increase the mobility of PV systems for quick response to disasters, automobiles and airplanes can be equipped with photovoltaic energy. This is a quick solution to transport people or supplies when there is no gasoline available in the aftermath of disasters and remote areas. This is also a solution to large-scale pollution created by so many gasoline-powered cars. Another potential benefit to the owner of solar-powered vehicles is the fact that having a solar fuel option reduces the risks associated with possible fuel shortages and price spikes. Even if gasoline supplies were eliminated, a solar power vehicle's owner could still maintain some degree of mobility. This concept may also have an appeal in other applications aside from personal transportation. The military may have an interest in an advance vehicle that relies only on solar for energy.

One of the solar-powered vehicles is a solar car—an electric vehicle that uses an electric motor for propulsion powered almost entirely by solar energy. The sun's energy is converted to electrical energy by photovoltaic cells which are then used to drive a solar car. A solar car differs from a conventional car in two ways: the type of engine and power source used. A conventional car uses a combustion engine that burns gasoline or diesel in cars and boats, and kerosene in seaplanes. The solar-powered cars use solar cells to generate electricity and electric motors for propulsion. These solar cells are usually mounted on the car's roof to have good exposure from the sun. However, the roof area of these cars, where the solar cells are to be placed is limited, therefore the amount of energy the cells can produce is small. In order to optimize the efficiency of solar-powered vehicles, special lightweight materials such as aluminum and carbon fiber are used. Photovoltaic cells can also be integrated into the body panels of a series hybrid vehicles, called vehicle integrated photovoltaic (VIPV). This represents a near-term opportunity for the widespread use of solar electricity for personal transportation. The solar-powered cars, however, lacks the basic safety equipment such as airbags, anti-lock brakes, and vehicle stability control systems as well as air conditioning and radio.

3.4.1   History of Solar Cars

In the United States, the first solar car was introduced in 1955 at the Chicago auto show. The car, called the Sunmobile, was only 12 inches long and was powered by 12 photovoltaic cells. The car had a 1.5 V motor which turned the driveshaft and transferred the energy to propel the wheels. Around 1962, a full size solar car was exhibited by a baker as a concept car but was never mass produced.

Since the 1970s, other individuals such as Ed Passerini (1977) and Larry Perkins (1982) constructed their own fully solar cars called the "Bluebird" and "Quiet Achiever" respectively which were exhibited in auto shows and competitions. In the mid-1980s, car manufacturers such as Mazda (Model Senku), Ford (Model Reflex) and Cadillac (Model Provoq) unveiled the solar hybrid concept: vehicles ranging from using solar power to run accessories to cells installed inside headlights.

In 2005, Mazda Senku used solar panels on its roof to help charge the battery. In 2006, Ford Reflex installed solar panels in the headlights while in 2008, Cadillac Provog used solar panels to power accessories such as the interior lights and the audio system.

In 2014, Toyota Prius used a rear solar panel located on the top of the rear part of the roof in order to power a ventilation system inside the vehicle. This solar panel roof can be activated with the remote entry to cool down the car before entering in to the vehicle. Drivers can also activate their ventilated solar panel roof before they exit their vehicle. Audi A8, a conventional gasoline-powered car, also uses solar panels for powering its air conditioning systems.

The first long-distance-driving car powered entirely by photovoltaic cells started in 1982 by Hans Tholstrup and Larry Perkins who performed the first manned journey across Australia from east to west in 20 days. The two pioneers covered this distance in a home-built car, the Quiet Achiever, which was an Australian-made British Petroleum-sponsored car. This achievement inspired solar car races—competitive races of electric cars powered by solar panels mounted on the surface of cars.

The first solar car race "Tour de Sol" was held in 1985 at Switzerland which led to several similar races in Australia, Europe, and United States annually from 1985 to 1993.

Currently two well-known car races are the World Solar Challenge and the North American Challenge which are contested by universities around the world and private entrepreneur teams. Both of these races require teams to build cars powered by not more than $6\,m^2$ of solar panels and race them over extremely long distances in a rough terrain.

The World Solar Challenge is a biennial solar-powered car race across the continent of Australia from Darwin, Northern Territory to Adelaide, South Australia. It covers a distance of 1877 miles (3021 km) in little over a week and is sponsored by the Bridgestone Corporation.

The American Solar Challenge, previously known as North American Solar Challenge, is also a biennial race which covers 1200–1800 miles across the country over several days. It is a competition for student teams from around the world to design, build and drive solar-powered cars and is held during the Summer in a long-distance road rally-style event. The routes are varied over the years and is designed to provide teams with an opportunity to demonstrate their cars under real-world driving conditions and test the reliability of onboard components. In 2014, the competition took place from Austin, Texas to Minneapolis, Minnesota covering 1700 miles (2700 km). The majority of these solar cars are developed for the purpose of racing because of the present challenges and limitations of using solar energy as a reliable source of vehicular power.

All of the above cars have used solar panels to power certain features of the automobile or use solar energy to power the vehicles for racing. Ford Inc. was the first manufacturer to develop the C-MAX solar Energi Concept car in 2014 with the potential to deliver the plug-in hybrid without depending on the electric grid or fuel. Ford's C-Max Solar Energi concept' car as shown in uses $16\,feet^2$ of high-efficiency rooftop solar panels and a large concentrating lens which speeds up the charging of battery for driving the automobile.

This increase in charging speed of the battery is achieved by placing a large oversized Fresnel lens on the stationary canopy above the car when parked and focusing the rays of light located on the car's roof. In order to maximize the incident sunlight, the car is programmed to automatically move forward or backward throughout the day, tracking the sun as it moves through the sky. It is estimated that with this concentrated lens, it takes the car about 6 or 7 hours to charge in an average US city's sunlight. The car can also be plugged to an electric plug-in to charge the car's battery at night or when there is not enough sun. According to a Ford Inc. estimate, the time taken to charge the car's battery using Fresnel lens is reduced to 7 hours in an average US city's sunlight. The C-Max Solar Energi will go 21 electric-only miles on a full charge and extend its range up to 620 miles on a gasoline engine. It is estimated that 75% of all trips made by an average driver are less than 21 miles which is provided by a solar powered battery. It is really a hybrid car which runs on electricity powered by solar for the first 21 miles and converts to gasoline after that. The car was developed by the collaboration of Ford, San Jose, Georgia Institute of Technology, Atlanta, and Californian-based SunPower Corporation.

In the case of popular electric cars like Chevy Volt, Nissan Leaf and Tesla Model S, the batteries of these electric vehicles can be charged by the electricity from one's own home's solar system. This can be achieved by using electrical vehicle charging station (EVSE) which uses 120V/15A circuit that plugs directly into a standard home outlet. These systems are provided by the battery-only electric vehicles and are considered to be trickle chargers. By investing in home's solar panels, the solar power not only saves in grid electricity but also in gasoline for clean environment.

3.4.2   Principles of Solar-Powered Car

A solar-powered car consists of the following components and its schematic is given in Fig. 3.9:

- Solar array
- Power tracker
- Batteries
- Motor controller
- Electric motor

Solar arrays are series of PV cells that collects the sun's energy and converts it to electrical current which will be stored in batteries to drive electric motor inside cars. Monocrystalline silicon PV cells are the most popular solar

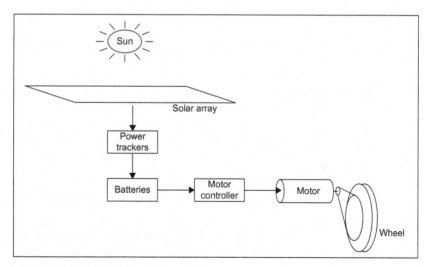

FIGURE 3.9   Schematic of a solar-powered car.

cells which provide the highest energy conversion efficiencies of all commercial solar cells. A typical silicon cell generates about 0.5–0.6 V DC under open circuit, no load conditions while the output current varies depending on the size (surface area) of the cell. In general, a commercially-available silicon cell produces between 28 and 35 mA/cm². The power of a PV cell depends on the intensity of the solar radiation, the surface area of PV cell, and its overall efficiency. As power is the product of voltage and current, larger modules will have larger output watt ratings. To increase the voltage output of a panel, PV cells should be wired in series and divided into several zones. In doing that, if one zone fails, the other zones are still producing power. If solar panels receive a limited amount of sunlight because of clouds or shading, their current will decrease. As a result, the panels will produce a lower amount of power. If a shaded cell is connected in series with other cells, the overall current of the series connection will be limited to that of the shaded cell. In extreme cases, this power imbalance can damage a solar panel.

The output voltage of the solar array does not need to match system's voltage if power trackers are used. Power trackers, which convert solar array voltages to the system voltage, are vital in solar cars. The power is automatically adjusted to match system voltage when the car drives under the shade allowing the system to run as efficiently as possible. The number of solar panels needed will depend upon the size and power requirement of the motor. It is recommended to have more power than is actually needed when it comes to using solar for motor operation especially if no battery back-up system is installed. Poor and intermittent fluctuations will result in motor performance if not enough power is provided.

A Power tracker converts the output of solar cell into proper voltage to be used by batteries to drive the motor. In the absence of a power tracker, the DC motor system would result in poor performance. In order to keep the output of the solar cell constant and high, maximum power point trackers are often used in most successful solar cars. This is achieved by using a chip (sensor) in the MPTT to continuously measure the output voltage of the solar cell, comparing it with the fixed battery voltage, and then determining the best voltage to charge the battery. This enables the energy from the solar cells to achieve efficiency greater than 92%. Motor controllers have become an essential electronic components of hybrid and electric cars by virtue of switching large amount of currents at high speed without any moving parts. A power tracker will also regulate the amount of electricity provided by the battery systems to prevent them from being damaged. The output of the power tracker is connected to the batteries where the power can be stored for the motor controller.

Batteries in solar cars are used instead of fuel tanks in a conventional car. The battery stores the incoming solar energy in a chemical form when the motor of the solar car is not running. Batteries provide stable power to the DC motors and as a result, the motors will work most efficiently as there are no interruptions in the amount of power supplied. The intermittent power fluctuations caused as a result of sunlight interference are overcome by a battery power supply. Power from the batteries is also needed to store energy for night time or cloudy day operation. Different type of batteries such as lead–acid battery, nickel cadmium (NiCd) battery, nickel metal hydride (NiMH) battery, lithium ion battery, and lithium polymer battery can be used for solar cars. The most commonly used batteries are made from lead–acid, lithium-ion, and nickel–cadmium. Common lead–acid batteries of the type used in the average family car are too heavy while others are lightweight and more efficient but are expensive. Solar-powered cars normally operate in batteries that range from 80 to 170 V.

The motor controller in solar-powered cars is required to regulate the amount of power supplied to the motor directly from the battery or solar array. Its functions include controlling the speed of the motor, direction of operation, acceleration, and regenerative braking. In order to maximize the efficiency of the motor, a controller will also condition the power before it reaches the motor. Three phase lines are used to power a DC brushless motor which is connected to the motor controller. The motor controller also has two serial interfaces. One of the interface connects to the motor and the other connects to a computer system and input controls. The input controls are used for throttling, forward and reverse direction toggling, braking, and various enabling switches. It is desirable that the controller should operate the motor with the highest possible efficiency under steady state-state operating conditions. Deviating from normal conditions, the controller should respond quickly to resolve the problem and resume normal operation to maintain a high level of energy efficiency.

Electric motors for driving solar cars are small, lightweight, highly efficient, reliable, and can also change the gears itself. The most popular choice for such a car is the permanent magnet brushless DC motors (BLDC) which has no brushes and requires extra electronic circuitry to perform the job of commutation. These motors can be made in many sizes and power ratings but are more complex and expensive than conventional DC brush motors. Because of the absence of brushes and mechanical commutation in BLDC, it has some additional advantages:

- Very high efficiency characteristics (98.2% recorded for an optimized Hallbach magnet arrangement) over a large power range
- High speed operation: can operate above 10,000 rpm under loaded and unloaded conditions
- High reliability: life expectations of over 10,000 hours meaning less overall downtime
- High power density, giving more torque per watt
- Responsive and quick acceleration: low rotor inertia allows them to accelerate, decelerate and reverse direction quickly
- Fast dynamic response because of high power density and torque to inertia ratio
- No arcing without brushes makes it vital when working in flammable gas locations

- Requires minimal maintenance
- Reduced noise and reduction of electromagnetic interference challenges
- Limited surface area of the car roof is one constraint on the panels' power production
- Cannot be tilted perpendicular to the sun for optimal energy capture unlike most photovoltaics on buildings or in solar farms, which either track the sun or are installed with a fixed southward tilt

3.4.3 Fully Solar-Powered Car

While most solar-powered cars have been made for racing, the first solar-powered family size car, Stella, made its debut in 2014 by traveling from Los Angles to San Francisco, United States, on Highway 1. Stella is a 4 seat, 4 wheel car that consists of a large solar panel that sits atop of the car's roof, as shown in Fig. 3.10.

The solar panel can power the car up to 500 miles on a single charge. Stella was designed and developed by a team of students from Eindhoven University of Technology and NXP semiconductors in 2013 to participate in the Michelin Cruiser Class of Bridgestone World Solar Challenge. The competition took place in the Australian desert where Stella completed 3000 miles journey between Darwin to Adelaide, and won first place.

Stella is the first solar-powered car with large solar panels on the roof. This car can communicate with other cars in a vehicle-to-vehicle communication (V2V) mode in order to exchange information with the vehicles ahead, allowing and adjusting to the most efficient driving speeds for ideal traffic flow. It can also communicate potentially with traffic police who could warn drivers about traffic jams and accidents. The car is fitted with a tablet screen which will tell the driver when the traffic lights will change. Stella is aerodynamically shaped, which helps in increased driving speeds by reducing frictional drag. The back of the car containing solar panels can be lifted to reach to the car's trunk. Other unique features of the car include a steering wheel which expands when the car is driven too fast and contracts when the car is driven too slow.

FIGURE 3.10 Stella solar-powered car. *Bart van Overbeeke Fotografie and Solar Team Eindhoven.*

Stella was designed by using a formal specification language with an associated toolset (mCRL2), which was developed at the Department of Mathematics and Computer Science of Technische Universiteit (TUe) Eindhoven in collaboration with LaQuSo, CWI and the University of Twente. The toolset supports a collection of tools for simulation, linearization, state-space generation, and to analyze specifications. The software tool was used for modeling, validation, and verification of concurrent systems and protocols.

Other specifications of Stella are given below:

- Can travel up to 500 miles (800 km) on a single charge at a speed of 80 miles/ hour (130 km)
- Powered by 1.5 kW solar panels on top of the roof
- Uses rechargeable Panasonic 13-kWh lithium-ion battery
- Measures 15 feet (4.6 m) in length and is less than 4 feet (1.2 m) tall
- Weighs 855 lbs (390 kg)
- Uses lightweight material such as carbon fiber

3.5   CASE STUDY OF SOLAR-POWERED AIRPLANE

The solar plane uses energy from the sun to power the electric motor rather than using internal combustion engines for propulsion of the plane. The plane consists of arrays of solar panels, rechargeable batteries, charge controller, an electric motor, propeller, and control panel for display as shown in Fig. 3.11.

Energy from the sun during the day, depending upon the intensity of light rays and their inclination, is converted into electrical energy. The plane employs rechargeable batteries to store excess energy to use it during the night or when there are clouds to power the plane. An MPTT is used to ensure that a continuous maximum amount of power is obtained from the PV cells on the solar panels. The power obtained from these cells is used first to power the propulsion system and onboard electronics and then charge the battery with surplus energy. When no power comes from the panels at night, only the battery supplies the power to various components to fly the plane.

The solar plane differs from a conventional plane in many ways. The solar plane does not have to land for refueling and can stay aloft as long as their batteries are charged. Solar-powered planes use propellers instead of jets to fly.

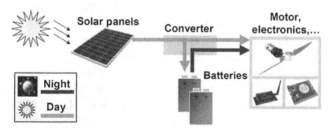

FIGURE 3.11   Schematic of a solar plane power transfer.

Some solar planes have up to 14 propellers. Solar planes are considerably light and travel at a much slower speed than conventional airplanes.

3.5.1 Power Received by Airplane

Ignoring the clouds, at midday, each square meter of land surface receives 1000 W or 1.3 horse power from the sun. For day and night (24 hours), this power averages out to be 250 W/m².

Considering that the PV cell has a surface area of 200 m², the maximum amount of power that the plane can theoretically capture from the sun is 250 W/m² × 200 m² = 50,000 W. Assuming the efficiency of solar cells to be around 20% and taking out the losses in the batteries and the electric motor which waste some power, the total efficiency of the propulsion chain is 12%.

Hence, the useful power at the plane's motor will be 12% of 50,000 = 6000 W or 8 horsepower. This is the energy from the solar panel to the propeller that plane is to use to fly day and night. This power will increase if the number of PV cells and their efficiency is increased.

3.5.2 History of Solar-Powered Plane

Solar aviation started with model aircrafts in the 1970s when affordable photovoltaic cells became available. The first flight of a solar-powered aircraft was designed by Robert J. Boucher of AstroFlight Inc., and flew in Nov. 1974, by launching a remote controlled Sunrise II from a ballistic catapult. David Williams and Fred To, two British nationals, were the first to attempt a manned flight in 1978 from Lasham Airfield, located in Alton, Hampshire, England.

Larry Mauro, a US national, flew a solar riser for the first time in Apr. 1979, at Flabob Airport, California, which was a solar version of his Easy Riser hang-glider. The plane flew for 800 m at an altitude of 1.5–5 m after charging the Ni–Cd battery with a 350 W solar panel.

Gunter Rochelet in Europe made his first flight in 1980 with Solair 1 using 2500 wing-mounted photovoltaic cells, generating up to 2.2 kW.

In 1981, Paul MacCready's team from United States developed a solar-powered experimental aircraft, the Gossamer Penguin, which crossed the English Channel in quick succession with a maximum power of 2.5 kW. This aircraft led the way to Solar Challenger and covered distance of several hundred kilometers which lasted several hours. The wingspan of this aircraft was 71 feet. (21.64 m), weighed 68 lbs, and used Astro-40 electric motor powered by a 541 W solar panel consisting of 3920 solar cells.

Eric Raymond, in 1990, used Sunseeker, a motorglider, to cross the United States in 21 stages in 121 flying hours over a period of nearly 2 months. The Sunseeker had a glide angle of 30 degrees and empty weight of about 196 lbs. It was equipped with amorphous silicon cells. An improved version was planned to become the world's first solar aircraft with two seats and is called the Sunseeker

Duo. The Sunseeker Duo with wingspan of 75 feet was driven by 20 kW direct drive motor powered by 72 lithium-polymer batteries connected in series. The batteries will store enough power to allow 20 minutes of full power and will have the capacity to cruise on direct solar power when above the clouds. The completion of Sunseeker Duo is currently supported by funds from kickstarter.

In the middle of 1990s several airplanes were built to participate in the "Berblinger" competition to honor the memory of the Tailor of Ulm (Germany) who attempted to fly hang glider across Danube in 1811. The aim of this competition was to fly an airplane with the help of at least 500 W/m^2 solar power which can climb up to an altitude of 1476 feet (450 m) and maintain horizontal flight.

3.5.3   Types of Solar-Powered Planes

There are two types of solar-powered planes.

- Unmanned aerial vehicles (UAV)
- Manned solar powered plane

An unmanned aerial vehicle has no pilot on board. These planes can be flown by a pilot at a ground station, remote controlled, or fly autonomously based on pre-programmed flight planes.

Unmanned solar planes include NASA's Helios (1999), the Zephyr (2010), Aurora Flight Sciences' Odysseus (2009) and SunLight Eagle (2012).

In 2005, founder of AC Propulsion, Alan Coccoid, flew a UAV named SoLong for 48 hours (day and night) nonstop entirely with solar power at El Mirage dry lake, California. The UAV had a wingspan of 5 m and flew at a higher altitude above the clouds with full sun and outside the air traffic control lanes with batteries powered by sunlight.

In 2010, QinetiQ (an Anglo-US company) made a nonstop flight of 336 hours and 22 minutes during 14 days of duration with its Zephyr drone. The plane flew at an altitude of 70,741.5 feet (21,562 m) with a wingspan of 39.37 feet (12 m) and weighed 27 kg (59.5248 lbs). The aircraft was launched by hand and flew by day on solar power which then used lithium-sulfur batteries charged by sunlight to power the aircraft at night.

A manned solar-powered plane was pioneered by Paul MacCready who flew Gossamer Penguin in 1980 with his son who handled the controls at the age of 13 in California. An improved version of Gossamer Penguin's flight Solar Challenger crossed the English Channel in 1981. The Solar Challenger flew at an altitude of 11,000 feet from Paris, France to an Air Force base in England covering a distance of 163 miles. The plane was powered by 16,128 solar cells attached to the plane's wings and became the first manned solar-powered flight. The purpose of these solar-powered flights supported by DuPont was to draw the world's attention to photovoltaic energy as a renewable and pollution free energy source for home and industry. This flight demonstrated the use of DuPont's advanced materials for lightweight structures.

3.5.4 Solar-Powered Manned Plane for Day and Night

Although the Solar Challenger discussed in the previous section is a manned solar-powered flight, it flew only short distances and did not fly during night. Solar impulse HB-SIA (referred to as Solar Impulse 1) is the first Swiss-registered solar plane which flew its first manned flight in Jul. 2010 for 26 hours including nearly 9 hours of night flying. Solar impulse 1 is a single seat monoplane capable of remaining in air up to 36 hours. Solar Impulse 1 was piloted by Piccard and Borschherg, who also completed flights from Switzerland to Madrid (Spain) and Rabat (Morocco) in 2012. This flight took them 19 hours and 8 minutes to complete a distance of 830 km. The two pilots also completed 6 stage, 5695 km bunny-hop flight from San Francisco to New York in 2013. The following are some of the specifications of Solar Impulse 1.

- Airplane length: 21,85 m (~71 feet)
- Height: 6.40 m (~20 feet)
- Wingspan: 63.40 m (208 feet)
- Airplane weight: 1600 kg (3527 lb)
- Average speed: 70 km/hour
- Solar cells: 116, 287,000 (wing: 10,748, Horizontal stabilizer: 800)
- Propulsion: 4 brushless sensor less electric motors
- Maximum cruising altitude: 8500 m (28,000 feet)

The purpose of launching a solar impulse HB-SIA prototype plane was to demonstrate the use of solar energy in powering an airplane that can fly around globe using pollution-free fuel. The lessons learned from more than 26 hours of nonstop flight are incorporated in the Solar Impulse HB-SIB, circling around the globe in 2015. The construction of Solar Impulse used MATLAB and Simulink software packages in various aspects of design, development, and mission planning. Simulink was used to create a system model of the airplane, its critical subsystems, and structural properties. A model of the airplane dynamics and flight simulator for pilot training was also produced and evaluated.

3.5.5 Solar Impulse HB-SB2 (Solar Impulse 2)

The upgrades version of solar impulse 1, namely, solar impulse 2 (Si2), flew from Abu Dhabi on March 9, 2015 (shown in Fig. 3.12) and reached Kalaeola, Hawaii, on April 21, 2016. This travel was covered in eight legs which included a 4-day and 21-hour leg between Japan and Hawaii. During this long flight, the plane suffered damage to the batteries and had to wait for 10 months for repair. The plane also had to wait for suitable weather and daylight in Northern hemisphere. The plane took off from Hawaii on April 21, 2016, and landed in San Francisco, CA, on April 24, 2016. The plane has since traveled from San Francisco to Phoenix, Phoenix to Tulsa, Tulsa to Dayton, and Dayton to Allentown, PA. Solar Impulse 2 took off from Allentown and landed in

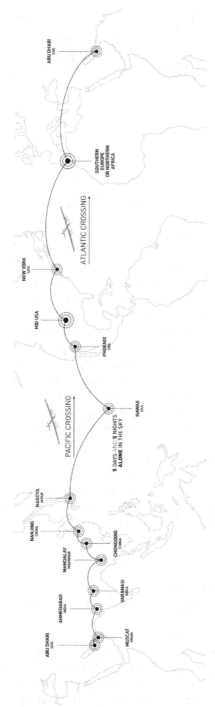

FIGURE 3.12 Route of Solar Impulse 2. *Solar Impulse.*

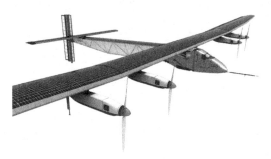

FIGURE 3.13   Solar Impulse 2.

New York on June 11, 2016, after spending some part of the journey turning around the Statue of Liberty. In the 15th leg, the plane took off from New York on 20 June and landed in Seville, Spain, on the morning of March 23, 2016, after traveling 6765 km in 71 h 8 min (2 days 23 h and 8 min). In the 16th leg, SI2 took off from Sevile on July 11, and reached Cairo, Egypt on July 13. It then took off on July 24, and landed in Abu Dhabi on July 26 on its final, 17th leg. Solar Impulse 2 covered a total distance of 43,000 km (26,719 miles) round the world in over 23 days of flight time without any fuel.

The flying is shared by Bertrand Piccard and Andre Borscherg of Switzerland, in a single seat plane (shown in Fig. 3.13). The plane is powered by 17,248 monocrystalline silicon cells of 135 μm thickness that are mounted on the wings and fuselage covering an area of 269.5 m². Solar Impulse 2 is made possible by 80 engineers, technicians, and several ground staff. Both of the pilots during the flight are connected by a satellite with Monaco Mission Control Center, where the flight' route, plane performance, energy requirements, and flight strategy is monitored and managed by a team of air traffic controllers, weathermen, planning engineers, and mathematicians.

The PV cells in SI2 collect up to 340 kWh of energy per day and have a high efficiency of 22.7%. These cells are highly reliable and do not lose power over time because of corrosion and breakage. The energy collected by the solar cells is stored in lithium polymer batteries which are housed in four engine nacelles. The energy density of the batteries is optimized to 260 Wh/kg and is insulated by high-density foam which is controlled by a system to control threshold and temperature.

The electricity produced by these high efficiency cells is enough for the plane to take off and recharge the batteries (on daily basis) that keeps the plane aloft through the night. Take off speeds of the plane are around 47 km/hour (29 miles/hour) and it will require 10 horsepower from each of four brushless electric motors (engines) connected to propellers, which is supplied by the PV cells and precharged lithium-ion batteries. After takeoff, the engines are quickly throttled back to seven horsepower to climb around 28,000 feet (8500 m) during the day to catch maximum energy from the sun. After the plane reaches this altitude, it will need around 2.5 horsepower per engine to cruise at the speed of approximately 80 km/hour. To conserve the power, the plane will slowly descend to about 5000 feet (1524 m) in the evening and night where it flies on battery power. These cycles will repeat until the next sunrise in the morning of the next day. The maximum speed of the plane at sea level is 90 km/hour and minimum speed is 45 km/hour. At maximum altitude, the highest speed is 140 km/hour and lowest speed is 45 km/hour. The plane switches between battery and solar power during the flight. During sunlight, the plane runs the propeller and charges the batteries and at night or in clouds, the propeller runs on the battery.

Four brushless sensorless motors, each generating 17.4 horsepower (13.5 kW), are mounted below the wings, and fitted with a reduction gear limiting the rotation speed of a 4 m diameter, two-bladed propeller to 525 rev/minute. The entire system is 94% efficient, setting a record for energy efficiency. The size of the plane's cockpit measures 3.8 m^3 and is not pressurized, heated or air conditioned. Due to the lack of heating in the cockpit the pilot is subjected to the changing temperature ranging from −40 to 40°C (−40 to 104°F). However, the cockpit is insulated to help the pilot to maintain a steady temperature.

An onboard computing system is used for the plane to selfcorrect and minimize its energy consumption. This is achieved by gathering and analyzing hundreds of flight management parameters, giving the pilot information to interpret to make decisions and transmitting the important data to the ground team. Most importantly, the energy consumption is minimized by providing the motors with optimal power for the specific flight configuration and battery charge/discharge status.

The body of the plane is strong and light. Its robustness is achieved by using a composite material of carbon fiber and honeycomb sandwich for the frame, with a strong fabric such as, Kevlar stretched across it. Rigidity and aerodynamic cross-section of the wing is obtained by constructing it with 140 carbon fibers spaced at 50 cm intervals. Solvay Inc. is Solar Impulse's main partner and technology partner, contributing to the project with technical support and its recognized competence in material development and applications. Following are the characteristics of Solar Impulse 2:

- Airplane length: 22.4 m (73.49 feet)
- Wingspan: 72 m (236 feet)
- Airplane weight: 4600 lbs (2086.52 kg)
- Average speed: 70 km/hour
- Solar cells: 17,000 spread across 270 m^2 of wings and fuselage
- Solar cells thickness: 135 μm
- Batteries energy density: 260 Wh/kg
- Number of batteries: 4
- Batteries weight: 2077 lbs (633 kg)
- Cockpit size: 3.8 m^3
- Propulsion: 4 electric motors
- Propeller size: 4 m in diameter

- Distance to cover: 35,000 km
- Flight time: 500 hours
- Maximum altitude: 8500 m (28,000 feet)
- Mission time: 5 months

3.5.6 Challenges of Solar Powered Manned Planes

1. Ground winds exceeding 10 mph (16 km) could crash the plane on the runway
2. Lack of turbulence layers will not allow the plane to ascend
3. Moving clouds and billows obstruct sunlight
4. Slow speed makes it easier to be swept by the wind
5. Higher energy is needed to cross oceans and continents
6. Being light as a feather makes it difficult to stabilize
7. Flying at an altitude of over 35,000 requires more power

BIBLIOGRAPHY

[1] American Solar Challenge. <http://americansolarchallenge.org/>, 2015.

[2] B. Brooks, Building a Solar Generator on Wheels. <http://tinyhouselistings.com/building-a-solar-generator-on-wheels/>, 2011.

[3] S. Buhr, The First Four-Seater, Solar-Powered Vehicle Hits the U.S. Road. <http://techcrunch.com/2014/09/24/the-first-four-seater-solar-powered-vehicle-hits-the-u-s-road/>, 2014.

[4] Bridgestone World Solar Challenge. <http://www.worldsolarchallenge.org/page/view_by_id/76>, 2015.

[5] Clinic In A Can: We Mobilize Medicine: Customized Containers for Global Health. Meeting Global Needs. <www.clinicinacan.org>.

[6] C. Conger, How Can Solar Panels Power a Car? <http://auto.howstuffworks.com/fuel-efficiency/vehicles/solar-cars2.htm>, Undated.

[7] Daily Mail Reporter, British Solar-Powered Unmanned Drone Finally Lands After Flying Non-Stop for Two Weeks. <http://www.dailymail.co.uk/sciencetech/article-1297165/Zephyr-British-solar-powered-unmanned-aircraft-finally-lands-flying-non-stop-weeks.html#ixzz3TdpqntFW>, 2010.

[8] Samsung opens Digital Village in South Africa, IT News Africa. <http://www.itnewsafrica.com/2014/04/samsung-opens-digital-village-in-south-africa/>, 2014.

[9] B. Ehrlich, The Lowdown on Mobile Photovoltaic Power Generators, This Article Originally Appeared on <www.buildinggreen.com> <http://greensource.construction.com/news/2010/100302Photovoltaic.asp>, 2010.

[10] M. Fay, Changemakers: Nomadic Communities Trust: Bringing Healthcare to Nomadic Populations in Remote Locations, Entered in Disruptive Innovations in Health and Health Care: Solutions People Want. <http://www.changemakers.com/disruptive/entries/nomadic-communities-trust-bringing-healthcare-nomadic>, 2007.

[11] From BIPV to Vehicle-Integrated Photovoltaics, Renewable Energy <World.com>. <http://www.renewableenergyworld.com/rea/news/article/2005/05/from-bipv-to-vehicle-integrated-photovoltaics-31149>, 2005.

[12] R. Goodier, SunBlazer Solar Trailers Could Electrify 7 Million Homes by 2020 Engineering for Change. <https://www.engineeringforchange.org/news/2012/11/01/sunblazer_solar_trailers_could_electrify_7_million_homes_by_2020.html>, 2012.

[13] C. Hiel, Achievement of 48 Hours Continuous Flight with a Solar-Powered UAV. Los Angeles SAMPE Chapter Composite Support & Solutions Inc. (Undated).

[14] The History of Solar Car Racing, Electric Vehicle. <http://www.solarcarchallenge.org/challenge/history.shtml>, 2015.

[15] K. Kathy Kowalenko, Lighting Up Haiti, IEEE volunteers help bring electricity to rural areas, The Institute, IEEE News Service (2011).

[16] L. Mearian, Ford Builds Solar-Powered Car. The Car Will Run at 100 mpg, Just Like Its Hybrid Predecessor, Ford Says. Computerworld. <http://www.computerworld.com/article/2487267/emerging-technology/ford-builds-solar-powered-car.html>, 2014.

[17] Mobile Power, Water and Communication, Three Crucial Systems in One Mobile Platform, Providence: <http://providencetrade.com/mobile-solar-and-wind-powered-energy-water-and-communications-system>.

[18] Mobile Solar Inc, The Power You Need Whenever You Need It. <http://www.mobilesolar-power.net/applications/>, 2015.

[19] A. Noth, R. Siegwart, Design of Solar-Powered Airplanes or Continuous Flight Given in the Framework of the ETHZ Lecture Aircraft and Spacecraft Systems: Design, Modeling and Control Partially Included of a Forthcoming Springer Book Chapter on Advances in Unmanned Aerial Vehicles, State of the Art and the Road to Autonomy Version 1.0, 2006.

[20] J. O'Callaghan, Meet Stella, the Solar-Powered Car That Drives 500 Miles on a SIGLE Charge-and Warns You When Traffic Lights Will Change. <http://www.dailymail.co.uk/sciencetech/article-2767806/Meet-Stella-solar-powered-car-drives-500-miles-SINGLE-charge-warns-traffic-lights-change.html#ixzz3SuNGhLDZ>, 2014.

[21] A. Parke, Solar Airplane Embarking on Mission Across the U.S. Ecopedia. <http://www.ecopedia.com/technology/solar-airplane-embarking-on-mission-across-u-s/#sthash.nRrunBKb.dpuf>, 2013.

[22] S. Parsons, Parsons, Solar-Powered Camel Clinics Carry Medicine Across the Desert. <http://inhabitat.com/solar-powered-camel-clinics-carry-medicine-across->, 2009.

[23] R. Podmore, R. Larsen, H. Louie, P. Dauenhauer, W. Gutschow, P. Lacourciere, et al., Affordable Energy Solutions For Developing Communities, April 2012 Show Issue. <www.ieeet-d.org>, 2012.

[24] Project, Solar Impulse Flight by Sun. Solaripedia. <http://www.solaripedia.com/13/170/1689/solar_impulse_airplane_solar_wing_panels.html>, 2009.

[25] S. Qazi, F. Qazi, Green technology for disaster relief and remote areas, in 121st American Society of Engineering Education Conference & Exposition, Indianapolis, IN, 2014.

[26] Refueling with Sunshine, Solar and Electric Vehicles, Solar Energy USA. Affordable Solar Solutions. <http://solarenergy-usa.com/solar-info/solar-and-electric-vehicles/>.

[27] Samsung Opens Digital Village in South Africa. Mobile and Telecoms, Top Stories, April 2, 2014. <http://www.itnewsafrica.com/2014/04/samsung-opens-digital-village-in-south-africa/>.

[28] S. Letendre, C. Herig, R. Perez, Real Solar Cars? Whitepaper on Vehicle Integrated PV, a Clean and Secure Fuel for Hybrid Electric Vehicles EV World, World of Electric Vehicles. <http://evworld.com/article.cfm?storyid=585&first=18101&end=18100>, Undated.

[29] Solar Trailers for Communications, Cameras and Lighting, NuSpectra Network Camera and Remote Communications. <http://www.nuspectra.com/mobile.php>, Undated.

[30] Solar-Powered Cars: How Do They Work? <https://www.yahoo.com/tech/solar-powered-cars-how-do-they-work-a72775157321.html>.

[31] "Solar Impulse: Around the World in a Solar Airplane," Solar Aviation. <http://info.solarimpulse.com/en/our-adventure/solar-aviation/#.VPe_zvldWn>.

[32] Stella Solar Car—mCRL2 201409.1 documentation <www.mcrl2.org/dev/user_manual/showcases/Stella_solar_car.html>.

[33] D.J. Unger, C-Max Solar Energi: Ford Goes Off-Grid with New Solar Car (+Video) <http://www.csmonitor.com/Environment/Energy-Voices/2014/0107/C-Max-Solar-Energi-Ford-goes-off-grid-with-new-solar-car-video>, 2014.

[34] World Health Organization, Harnessing Africa's untapped solar energy potential for health, Bulletin of the World Health Organization. <http://dx.doi.org/10.2471/BLT.14.020214>, 2014.

[35] World's Largest Solar-Powered Hospital Opens in Haiti. <http://www.designboom.com/architecture/worlds-largest-solar-powered-hospital-opens-in-haiti/>, 2013.

[36] W. Young, Photovoltaic in disaster management, SATIS 2001 Conference, Kingston, Jamaica, 2001.

[37] W. Young, Photovoltaic Applications for Disasters Relief. FSEC-CR-849-95, University of Central Florida Solar Energy Center, 1995.

[38] K. Zipp, Solar Makes Remote Medical Care Possible. <http://www.solarpowerworldonline.com/2012/09/solar-makes-remote-medical-care-possible/>, 2012.

[39] K. Zipp, Clinic In A Can: First Fully Solar Medical Facility, Wichita nonprofit shipping two portable medical clinics overseas (2012).

Chapter 4

Portable Standalone PV Systems for Disaster Relief and Remote Areas

4.1 INTRODUCTION

Portable standalone PV systems are small, easy to transport, and are commonly used to provide electricity in the aftermath of a disaster or provide services to remote areas in the absence of an electric grid. These systems may be as small as a flashlight or as big as a suitcase for water filtration. Each complete system consists of a basic set of components: a solar panel, solar charge controller, inverter, battery, wiring, a carrying case or a backpack and sources for 12-V DC and/or 120-V AC power. The components of these systems are chosen to match each other's functionality. The solar charge controller is chosen to match and optimize the performance of the solar panel and the battery. Solar panels in these systems are waterproof while other components such as inverters and electronic devices are not. As a result, care should be taken when the system is exposed to water, rain, snow, or saltwater.

Many manufacturers have developed portable, lightweight hand-carried and reliable solar-powered systems to provide electricity in the aftermath of disasters, or for limited energy in remote areas of the world experienced by three billion people. Such a device can also be used by people in developed countries who are dealing with power outages, for camping or for an outdoor party. The goal of each system is to provide power for portable electronic devices in the absence of the electric grid. The more power and stored energy one requires, the larger and heavier the system becomes.

4.2 FEATURES OF PORTABLE SOLAR SYSTEMS

1. *Lightweight and Easy to Carry*
   Should be lightweight and easy to carry, especially for disaster relief or outside activities in remote areas.

Standalone Photovoltaic (PV) Systems for Disaster Relief and Remote Areas.
DOI: http://dx.doi.org/10.1016/B978-0-12-803022-6.00004-6

2. *Length of Exposure*

Should provide power for a longer period of time with shorter time of exposure to the sun. Longer exposure time will require longer waiting time for the device to become fully functional. During disasters and outdoor activities in remote areas, time cannot be wasted.

3. *Deliver power in remote areas*

Should have the option to be charged from the electrical home outlet as the remote location may have hindrance to sun or no sun.

4. *Capable of powering multiple devices*

Many mobile devices including laptops need to be charged. The solar charging system should have capacity of charging these devices for more than 3 hours and should be compatible with different energy outputs.

5. *Capable of charging and delivering power in all weather conditions*

This requires the portable system to be weatherproof as the majority of the applications of this device would be used outdoors and in remote areas.

6. *Semiflexible solar panels*

Semiflexible solar panels can be folded or rolled up for easy transport. They can also open up to provide a greater surface area than many rigid panels.

7. *Appropriate solar panel attachment*

Suitable attachment option to securely attach to a tent, backpack, bike or boat for transport.

4.3 TYPES OF PORTABLE PV SYSTEMS

A portable standalone PV system is a combination of components and raw materials that, when integrated into a system, will produce electricity and store energy from sunlight. These systems have been used to provide electrical power for disaster relief, remote homes, emergency power, and camping where no other source of electricity is available. Portable PV systems can be divided in the following categories based on the size and method of transport:

- Compact portable PV systems
- Solar backpack systems
- Solar suitcase/briefcase systems
- Foldable PV systems

4.3.1 Compact Portable PV Systems

Compact portable solar systems are low powered, compact, lightweight, and capable of powering electronic devices or providing light in remote areas. A simple case is a solar lantern, a flashlight and a cell phone charger that can be carried by hand, in a coat pocket, a briefcase or a backpack. A smaller PV system, although easier to pack, takes longer to charge a battery. New systems are using flexible PV modules (panels) mounted on jackets, briefcases, backpacks, purses, and other smaller surfaces. Even the smallest versions include adapters to charge multiple electronic devices.

4.3.1.1 Solar Lantern

A solar lantern (Lighthouse 250) was developed by Goal Zero to provide light in the aftermath of disasters around the world including the 2010 earthquake in Haiti, 2011 tsunami in Japan, 2012 hurricane Sandy in northeastern United States and the 2013 Typhoon Haiyan in southeast Asia. Goal Zero was formed in 2008 after staff worked and gained experience in humanitarian work in the Democratic Republic of Congo. Lighthouse 250 as shown in Fig. 4.1A consist of two 3 W LEDs and a built-in 4.4 Ah (Li-NMC) rechargeable battery which produces 250 Lumens of energy and is adjustable. The battery can be fully charged from any of the external solar panels available from Ground Zero, a USB power source or a foldable inbuilt crank. The lantern takes 7 hours of charging from the sun to run up to 48 hours on low intensity setting and 2 hours 30 minute on high intensity setting. The lantern can also be used to charge cell phones and power electronic devices by connecting them to USB port by way of built-in cord.

(A)　　　　　　(B)

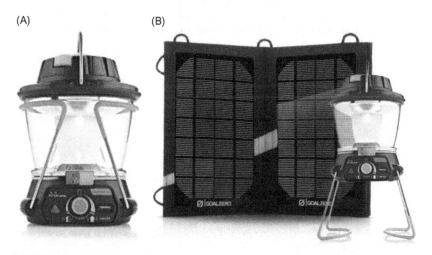

FIGURE 4.1 (A) Solar Lantern; (B) solar lantern and charging foldable solar panel.

In the absence of a power source or sun, 1 minute of built-in cranking can provide 10 minutes of run time. The solar lantern is capable of spraying light both in a full 360 degrees and directional in 180 degrees. The lantern is equipped with emergency flashing of red LEDs, which indicates when the battery needs its next charge. The portable lantern weighs 1.1 lbs and is supported by two foldaway rubberized legs. The Lighthouse 250 costs around US$80 and part of money from the cost is used to send these lamps to poor communities of the world with no access to electricity. The lantern is capable of charging small USB-powered electronic devices and supplying power to USB and 12 V devices, making it suitable for emergency and outdoor usage. The Goal Zero Lighthouse 250 kit uses Goal Zero Nomad 7 solar foldable panel to charge the lantern in the sun as shown in Fig. 4.1B.

Goal Zero Nomad 7 consist of two monocrystalline solar panels with capacity of 7W (maximum current 0.5A) capable of charging handheld USB/12V devices directly from the sun. It is an ultralight system weighing only 0.8 lbs and measures 6 × 9 × 1 inches. The solar panel is foldable, rugged and is weather resistant. It does not have an internal battery and contains built-in pockets to protect devices or cables. In addition to charging solar lantern, it can be used to charge most cell phones, smartphones, and GPS devices.

4.3.1.2 MPOWERD Luci EMRG Solar Lantern

An inflatable solar lantern which can be collapsed for easy transport is manufactured and marketed by MPOWERD Inc. It consists of solar cells, a built-in lithium-ion battery, and ten white LEDs providing enough light to illuminate a 10-foot room for 6–12 hours depending on the setting. The lantern as shown in Fig. 4.2A–C looks like a translucent cylindrical beach ball, has a 5-inch diameter, measures 4 inches in height when inflated, and 1 inch when collapsed. The underside solar cell will recharge the battery in 6 hours of sun or artificial light and provide light for 6 hours at bright setting or 12 hours at dimmest setting.

(A)   (B)   (C)

FIGURE 4.2 (A) Luci EMRG by MPOWERD inflatable solar lantern; (B) lantern top side with solar panel; (C) collapsible view.

The MPOWERED Luci EMRG solar lantern costs only US$20, but for every lantern purchased the manufacturer (MPOWERD Inc.) will donate another one to the communities in need around the world. Since the start of the "Give Luci initiative" for solar justice, 500 lanterns have been sent to storm-affected families in the Philippines, 300 to girls living in refugee camps, 500 to single women households in Africa and 500 to needy communities in Amazon. In a recent report by GlobeScan (http://www.globescan.com/), nearly all the Haitians using kerosene based lamps for lighting were replaced with the received Luci inflatable solar lantern and 98% of families reported a decline in both breathing problems and eye irritation.

4.3.1.3   Ultralight Solar Charger and Panel Power Module

The Ultralight solar charger and panel power module is a portable system weighing only 3.6 ounces. It is used to charge USB-cabled devices such as cell phones, GPS devices, and Ipods from the sun in the field or through lithium-ion battery by night or on cloudy days. Plug-in via USB or adapters for the battery charging includes various types of cell phones from Ericsson, Motorola, Nokia, Samsung and others. As shown in Fig. 4.3, this solar charger and power module consists of a high efficiency solar panel with 5.5 V/320 mAh output and maximum charging unit of 1.76 W.

FIGURE 4.3   Ultra-light portable solar charger.

The solar panel measures 6.5 inch × 4.875 inches and consist of USB female output port. The lithium-ion battery output is 5.5 V/500 mAh with the maximum capacity of 1.0 mAh. The charging time for battery directly from the sun is 3–4 hours and by PC USB is 2 hours. The kit contains an adjustable stand for the solar panel to prop up to its optimum angle as well as a carrying case with integrated clips and hook and loop straps to connect to the backpack.

4.3.1.4   SolarWrap

Another portable, compact and foldable item called the PowerSync SolarWrap is developed by Bushnell Inc. to collect and store power. The power produced can charge any electronic devices with USB input such as a cell phone or digital camera. The SolarWrap uses durable flexible solar panels that roll up into an integrated protective case as shown in Fig. 4.4A. Each solar cell on the panel is wired independently so that a break in one solar cell will not affect the rest of the solar cells. The SolarWrap's roll-out design comes in three versions: the mini 100, weighing 3.1 oz; the 250 weighing 9.2 oz; and the 400, weighing 10.1 oz. SolarWrap mini 100 measures 4.3 × 1.25 inches and its roll-out length is 18.25 inches. The SolarWrap 250 measures 9.125 × 2.4 inches and its roll-out length is 17 inches, while the SolarWrap 400 measures 9.125× 2.4 inches and its roll-out length is 29.25 inches.

All three paper-thin solar panel rolls up into a protective case measuring the given sizes. The charging time from the wall socket for all the three SolarWrap is 4 hours but their charging time from the sun is different because of their varying sizes. The SolarWrap mini takes 10 hours to charge from the sun while SolarWrap 250 takes 6 hours and SolarWrap takes 3.5 hours to charge. Although it takes a longer time to be charged from the sun, a phone can be charged from 50% to full charge in 1 hour. The power output for all the three versions is 5 V/1 A and all of them use 2.2 Ah lithium- ion batteries with one USB outlet.

This paper-thin rolled out SolarWrap can be positioned anywhere where there is sunlight including on the hood of a car, a tree, a backpack or a wall where these devices can be anchored to remain flat. Charging from the battery can be accomplished without the risk of damaging the panel as the USB outlets are on the ends under the caps. The SolarWrap comes with both a standard and mini USB attachment. It also includes a cord with micro- and standard USB ends in addition to an adapter for charging the SolarWrap from an outlet as shown in Fig. 4.4B and C. While in the sun, SolarWrap can charge a phone and its battery simultaneously; this means that the SolarWrap's battery can be used even when there is no sunlight. Despite its many advantages, it has a few disadvantages such as the battery is small, the unit is not waterproof, the unit is not very durable and has a limited battery life.

(A)   (B)   (C)

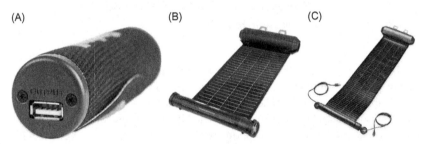

FIGURE 4.4   (A) Solar panel roll up; (B) solar roll out; (C) solar panel roll out with input and output cables.

4.3.2   Solar Backpack Systems

A backpack is a convenient way of carrying things without using one's hands. It can also be used for generating electricity for people on the move. A typical backpack as shown in Fig. 4.5A and B consists of the outer surface which is used to attach a solar module. It also has interior space to include a storage battery and other components when not used for energizing electronic devices. Other components of the backpack include a flexible monocrystalline or thin

(A)   (B)

FIGURE 4.5   (A) Solar backpack; (B) Larger solar backpack.

film solar panels, batteries, light bulbs, charge controller, connecting wires and plugs. NASA and the US Army have used copper indium gallium diselenide (CIGS) solar cell for its unbreakable strength and flexibility. Since the backpack has a limited surface area to attach a solar module, it is limited to smaller wattage. The larger the solar panel, the more sunlight it collects and the faster it gets converted to power stored in a battery. Depending on the type of solar module, a backpack can charge an electronic battery of 3.7V, 1.02Ah lithium batteries in 2–3 hours. The applications of solar backpack include powering a GPS system, a travel lamp, a digital camera, a palm pilot, and other rechargeable electronic devices.

The US army first used the solar backpack for communication equipment in 2010 in Afghanistan known as REPPS (Rucksack Enhanced Portable Power System). A similar system was used by the US Air Force and Marines in 2009 and earlier. The US Air Force used solar panels on shipping containers while the US Marines developed suitcase units of foldable solar panels which could be carried as a backpack. REPPS is a 62-W foldable solar panel blanket, weighing around 10lbs making it easier to be carried as a backpack. The charging time for REPPS is only 6hours and after that it could be used to power electronic devices. It can also convert generated DC into AC voltage. In the absence of sun, REPPS can be charged from other power sources such as wall outlets, disposable used batteries, and vehicles. The University of Massachusetts at Dartmouth and US Army are currently working to develop photovoltaic fabrics which could be integrated into backpacks, tents (shown in Fig. 4.6) and clothing through an industrial partner with expertise in flexible solar power.

Because of ease in implementing a power system in the field and concern for expense and risks in transporting fossil fuels for troops in the remote bases, the US Military has chosen solar energy as a fuel for the field operation.

FIGURE 4.6 University of Massachusetts at Dartmouth and US Army solar tent.

4.3.3 Solar Suitcase/Briefcase Systems

The solar suitcase or a briefcase is a portable solar power system that can accommodate one or more solar panels to provide power for mobile communications equipment, medical instrumentation, and computers to meet the power need of disasters and remote areas. These systems usually consist of one or more high efficiency solar panels, batteries, inverters, lights, connecting wires and cables and medical devices. Larger and multiple solar panels increase the surface area of the panel which collect more light to get the most out of solar panels facing the sun the right way.

Aspect Solar have developed the SunSocket system with solar panels that can track the system as it moves across the sky, hence boosting the efficiency. It is the first portable solar generator that utilizes sun tracking technology normally used in large solar power stations and larger home installations to increase the efficiency in conditions such as low-angled low-intensity light in Winter or cloud cover. The sun tracking technology in the SunSocket generator as shown in Fig. 4.7 allows the solar panels to rotate automatically toward the sun throughout the day which helps to increase the efficiency of the generator up to 30% without repositioning the solar panels.

The SunSocket solar tracking system consists of three monocrystalline solar panels of 60 W, 250 W battery, a 100 W inverter with a universal plug in addition to USB plugs, a 12 V plug, charge meter, and a port for charging the battery from the AC mains all housed in a suitcase. The inside solar panel is expanded and slid into a 44.8-inches-wide panel which begins to track the sun allowing it to charge the battery.

The solar generator stores 250 Wh and takes 5 hours to charge the battery in good sunlight and 9 hours in poor sunlight. It uses a lithium iron phosphate (LiFePO$_4$) battery which is much lighter than lead–acid battery commonly used in other portable solar generators. The LiFePO$_4$ battery has longer life span than other lithium ion batteries but is more expensive than lead–acid

FIGURE 4.7   Portable solar tracking system in a suitcase.

batteries. A fully-charged LiFePO$_4$ battery offers 25 hours of charging time for rechargeable devices. The combined weight of the SunSocket generator consisting of solar panels, 250 W LiFePO$_4$ battery and hard metal case is only 25 lbs, and it folds into $16.5 \times 20.2 \times 4.2$ inches case.

The portable solar system is used to power a wide range of electronic devices from laptops, phones, tablets, to illuminate a room or a campsite. It is an important power source for emergencies and power outages as well as for remote areas in the aftermath of disasters. The system is self-contained where all of the necessary output sockets are built inside the unit to be used for charging any portable devices simultaneously. In the absence of sun, the LiFePO$_4$ battery can be charged from the wall outlet. Despite its many advantages over the other portable solar generators, it is not very rugged and is not waterproof, making it best for use only in dry weather.

4.3.3.1   Solar Suitcase for Water Purification in Disaster and Remote Areas

Clean water is the foremost requirement for sustaining life in the aftermath of a natural disaster and in remote areas. According to a UN Water report, 783 million people, mostly in the developing countries, do not have access to clean water and approximately 8 million people die each year from diseases related to water contamination. Trunz Water Systems has developed a portable water purification system called Trunz Mobile System Survivor 300 (shown in Fig. 4.8), which can be powered by the sun or 12 V DC from an automobile.

Trunz Mobile System Survivor 300 consists of a foldable solar panel, a prefilter, an ultrafiltration membrane, an activated carbon filter and a diaphragm feed water pump. The water purification system is based on ultrafiltration technology which is a type of membrane filtration in which forces like hydrostatic pressure forces a liquid against a semipermeable membrane. A semipermeable membrane is a thin layer of filtering material which separates substances when a driving force is applied across the membrane. This filtering method is suitable for purification of fresh water sources from rivers, wells, lakes, or boreholes.

Trunz Mobile System - Survivor 300

FIGURE 4.8 Solar suitcase for water purification.

Survivor 300 is capable of delivering approximately 50 gallons (180 L) of fresh water per hour. It uses a 135 W 12 V foldable solar panel or 12 V DC source which powers the feed water pump to filter fresh water under pressure through a high-technology hollow fiber membrane, retaining all the natural minerals in the water. The diaphragm feed pump creates 43 psi (3 bar) pressure which produces maximum throughput of 66 gallons per hour (250 L/hour) of water completely free of viruses and bacteria. Its turbidity for organic contamination and particles is less than 5 NTU that makes the water drinkable for humans. Survivor 300 uses an activated carbon filter of 10 inches in 10 micron and a pre-filter of 175 micron. The deployed membrane removes viruses, bacteria, cyst, organic contaminants and others down to 0.04 μm.

The filtering system measures 33.5 (L) × 21.7 (W) × 13.8 (H) inches and weighs 113 lbs including batteries and solar panel, making it portable and easy to transport. It can be air-lifted in small airplanes or a small automobile and can be redeployed quickly. Trunz Mobile System Survivor 300 can be put into operation in less than an hour which makes it easier to deploy at different locations instead of moving the refugee(s) to a central location.

4.3.3.2 Portable and Rechargeable Solar Suitcase for Emergency Survival

LinorTrek of North Carolina, United States developed and assembled a solar briefcase called OTG15W for emergency use as well as for providing reliable power at work, home or off-grid in remote locations where power is not available. As shown in Fig. 4.9, The solar suitcase consists of a built-in 15 W silicon monocrystalline panel, built-in inverter, charge controller, 3 W LED light, two

(A)   (B)

FIGURE 4.9   (A) Front and (B) side view of solar suitcase for Emergency Survival.

12 V DC/9 Ah SLA batteries and LCD display panel to check battery status and charging rate. It also contains a master battery cut-off switch to ensure long battery life as well as low battery warning and overload protection. Another feature includes programmable mode which turns the LEDs and other devices on at dusk. It also has a sleep timer to automatically turn power off or on.

The 15 W solar panel recharges 12 V/18 Ah battery, and provides power outputs for 12 V DC, 120 V AC, and USB. OTG15W solar suitcase can also be fully charged from an AC outlet. The charging current for full sunlight is approximately 750 mA and charging current from AC is approximately 2 A. The AC input is 120–240 V/50–60 Hz and AC output is 120 V, 150 W at 60 Hz modified sine wave. The DC output is 12 V/8 A and USB output is 5 V/500 mA. The suitcase is equipped with 12 V DC outlet (fused at 8 A) for powering DC appliances using cigarette plug and supports a 500 mA USB power port for powering USB appliances. Many electronic and electrical devices can be easily charged using power outputs for 12 V DC, USB, and 120 V.

The LinorTrek OTG15W solar suitcase can be used to operate a laptop, charge a cell phone, power lights, radios or power tools in the aftermath of disasters and remote areas. The built-in 3 W LED light is super bright and lights up a room or a campsite. It was successfully used to run laptops and charge cell phones for 3 days during the Feb. 2014 ice storm in eastern Pennsylvania, where more than 700,000 people at one point had no electricity. This solar briefcase is a great survival tool in the aftermath of disasters and power outages when communications and power to run electronic and electrical devices are critical. For nonemergency situations it can be used for home electronics with standard outputs for AC and DC TVs, CD players, lights, computer, cash registers and small appliances. It can also be used to power 12 V car electronics, iPads, DVD players, cell phones, tablet computers, and other portable electronic all at the same time. The suitcase measures 19.5 (W) × 3.5 (D) × 12.5 (H) inches and weighs approximately 21.5 lbs, which makes it portable and self-contained so that it can be transported anywhere.

4.3.3.3 Solar Briefcase

Sunshine solar has developed two portable solar briefcases that can be used to provide power in the aftermath of a disaster or for remote areas. Their applications include powering a DC fridge, laptop computers, electric hand tools, GPS and cell phones. Both the suitcases measure about 13× 20× 1.5 inches and weighs 9.25 lb as shown in Fig. 4.10. Both the suitcases consist of adjustable support panel which helps to maximize power output. They also include male and female cigarette connections, a blinking charge indicator and a blocking diode to prevent reverse charging fitted with s-s connector. When not in use, their briefcase configuration provides excellent protection.

(A)   (B)

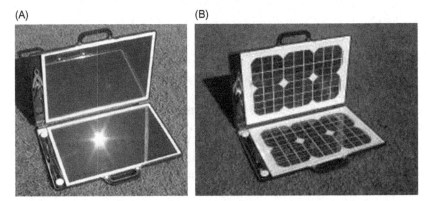

FIGURE 4.10 (A) Solar briefcase using amorphous; (B) solar suitcase using monocrystalline panels.

The first solar briefcase shown in Fig. 4.10A uses an amorphous solar panel with power output of 13 W. The 12 V/7 A batteries are recommended for storage. The second briefcase shown in Fig. 4.10B uses a monocrystalline solar panel with a peak power output of 35 W. It has peak working voltage of 17.5 V and peak working current of 2 A.

4.3.4 Foldable PV Systems

Foldable PV systems are a relatively new addition to solar power systems that are available. Most of these systems are based on the use of flexible thin film solar panels which are thinner and lighter than traditional PV panels and can be easily transported to be used anywhere. These foldable modules can be made rollable which allows them to be used in places where traditional solar modules are impractical. Unfolding these modules exposes them to greater surface area while still being something that can be rolled up for portability. Most of these systems weigh less than 1 lb but can output plenty of power when given good exposure to the sun. Foldable PV systems are not easily damaged than some of the rigid types of solar charging systems.

Foldable PV modules in the range of 5–10 W are often used to recharge cell phones, media players and e-readers. The 25–30 W modules are suited to power small laptop computers, tablets and RV batteries in addition to recharging iPads and cell phones. These nonglass modules can be folded up to the size of small book and easily stored in a backpack or a car glove box. Most of the fold-up modules are used to recharge batteries in devices that can accept 12 V DC as they do not include storage batteries themselves. A portable 12-V DC module with a 25 W can provide more than 2 A of output in full sun. Foldable PV modules of over 60 W are available which can be used to power radios, satellite communication systems in remote locations and laptop computers during camping or hiking as shown in Fig. 4.11.

FIGURE 4.11 Powerenz foldable system for powering laptop and cell phone during a camping trip.

A built-in metal grommet on each corner of foldable module is often used in order to suspend the foldable modules between two trees, hang them on the side of the wall or tie them on the hood of the car while camping. Flexible foldable solar panels are lightweight, portable and easy to use in an emergency for rescue workers who are on foot and backpacking in the field. Foldable flexible solar panels usually use thin film technology to power small electronic devices such as a battery charger, cell phone, or mp3 player. Solar panels made with CIGS are proven to provide higher efficiency than other flexible solar cell technologies including amorphous silicon. Some manufacturers such as Powerenz, Inc. are now offering lightweight 150–300 W monocrystalline foldable panels with over 24% efficiency.

4.3.4.1   Powerenz LP40 Waterproof Portable Solar System

Powerenz Inc. developed a portable power system (LFP 40) using foldable solar panels that can be used for powering a variety of devices that require 12 V DC up to 15 A/180 W, and 120 V AC up to 320 W. LFP40 is best suited for charging cell phones, laptop computers, communication radios, night vision systems, lighting fixtures, TVs, GPS, small refrigeration units, satellite phones and repeaters, flashlights and small power tools. It consist of a waterproof and crushproof hard case which houses battery assembly, solar charge controller, power inverter,

AC-powered battery charger, small battery charger, wiring, extension cable, and accessories. The foldable solar panels, shown in Fig. 4.12A and B, are not housed in the hard case and can be recharged by the sunlight.

(A)   (B)

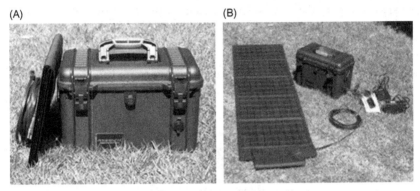

FIGURE 4.12   (A) LPG 40 Foldable solar system in a hard case (B) components.

The details of the important components of the foldable solar system are given below.

4.3.4.1.1 Solar Panel

To increase the surface area of solar panel a foldable 75 W/3.75 A panel with an efficiency of 16–17% is used to charge the battery. The solar panel requires 8–9 hours of peak sunlight to fully recharge the battery from complete depletion. The solar panel weighs 8.3 lbs and measures 18.83 × 12 × 1 inches folded and 63.75 × 18.83 inches when it is unfolded.

4.3.4.1.2 Battery

To make the system portable, a lithium iron phosphate (LFP) rechargeable battery of 12.8 V/40 Ah is used which weighs 40–50% less than lead–acid batteries of comparable charge capacity. A LFP battery provides 2000 days of charge/discharge cycles before losing its rated charge capacity compared to 250–300 days for Gel lead–acid or AGM batteries. LFP's battery stores 512 Wh of energy and can be charged directly from the solar panel using solar charge controller via a cable. It can also be charged from 10 A, 100–240 VAC powered charger through the wall outlet. The AC recharging time is 4.5–5.5 hours from when it is completely depleted.

4.3.4.1.3 Solar Charge Controller

A solar charge controller is used to keep the battery from overcharging by regulating the voltage and current coming from the solar panel to the battery. It is programmed at 15-A/200-W unit and uses MPPT (maximum power point tracking) to accelerate solar charging of the battery up to 30% per day. MPPT checks the output of the solar panel compares it to the battery voltage and adjusts it to the best voltage in order to get maximum current in to the battery. The solar charge controller uses a 25-A circuit breaker to protect it against overcurrent and has a baseline of continuous power consumption of 35 mA. The 12-V DC power is accessible via a female cigarette lighter socket that is implanted in a sidewall of the hard case, and is protected against overcurrent by the internal circuitry of the solar charge controller.

4.3.4.1.4   Power Inverter

A power converter changes the 400 W 12 V DC output of solar panel into 120 V modified sine wave AC unit with 90% efficiency. The unit has 2 AC receptacles and is protected internally against overcurrent. The inverter is connected directly into two terminals of battery via a pair of alligator clamps. The solar power input connector is implanted into a sidewall of the hard case.

The 12 V DC battery can be charged from AC wall outlet in case there is no sunlight. This is achieved by using 10A 110–240 V AC voltages at 50 or 60 Hz that plugs directly into the AC-powered battery charging port. The solar input connector is embedded into a sidewall of the hard case. The LPG 40 in hard case weighs 42 lbs and is ATA 300, IP67, STANAG 4280, and DS 81–41 approved for being waterproof, dustproof, and crushproof. It is black in color and can be carried easily by a rugged handle.

Other versions of LPF 40 solar portable system are also available in a sling pack and a hard case in leafy woodland camouflage.

4.4   CASE STUDY: A PORTABLE SYSTEM FOR DISASTER RELIEF AND REMOTE AREAS

Disasters cause human suffering and create human needs that victims cannot alleviate without assistance. Disaster affects more than one person and may impact the safety and security of an entire community, overwhelming local resources. Most of the communities are not well prepared for disasters and disruption of services in its aftermath. Electrical power is one of the first things to be disrupted and might take time to be restored or bring other clean sources of power in the absence of electricity from the grid. New Vision Renewable Energy (NVRE), a Christian community organization, has been working since 2009 to provide safe lights and clean water to communities in need and also in the aftermath of a disaster. The organization evolved out of a friendship between an independent-minded renewable energy innovator and a Christian Community Development leader to reenergize communities and to help people to help themselves.

NVRE is a nonprofit organization, located at Chestnut Ridge (shown in Fig. 4.13A) in North Central West Virginia that works locally and globally with

(A)   (B)

FIGURE 4.13   (A) NVRE organization and volunteers; (B) NVRE light for school and community. *NVRE.*

faith-based and community organizations to get renewable products into the hands of their members and partners at affordable rate. They also provide training to the needy communities to produce their own renewable energy systems.

NVRE developed three portable systems for light, water filtration, and power to the people in remote areas and for the disasters. These systems are easy to carry, easy to deploy, that can be used to help those affected by a disaster and return to a normal life in a short time.

4.4.1   Light for Schools and Community

The first portable unit shown in Fig. 4.13B is a system for light and life. In the absence of electrical grid many students in the developing world use kerosene lamp as a source of light. These lamps are dangerous and it is estimated that more than 4000 people die each day by fire, severe respiratory and vision problems. To reduce the number of death and provide safe lights, NVRE is engaged with partners to bring clean light to students and teachers throughout the world. NVRE is also working to provide safer and cleaner light to orphanages and churches through portable solar systems. In the absence of proper lighting in the churches, it is difficult for families to worship together or for children to take part in educational and recreational activities. It is risky for the families to walk because of the fear of falling and being attacked by predatory animals or robbers in the dark. In schools and orphanages children cannot finish their homework in time and as a result fall behind in education and hence opportunities for better living. New Vision has donated over 1200 lights in 28 countries around the world. This is done by finding a partner in developing countries and raising funds in United States through Kickstarter and fundraising events. The details of the light for schools and community are given below:

- Solar panel: 10 W polycrystalline
- Battery: 12 V/9.8 Ah lithium polymer
- Light sources: Three super bright LED light strips, ultra lightweight design

The unit requires 1.5 hours of charging providing 1 hour of battery life. It weighs 4.5 lb and measures 12 × 7 × 2.75 inches and is equipped with one USB port which can be used to charge the cell phone. The cost of the unit is US$100 and is available as a kit or as an assembled system. Part of the cost is used to send the units to the people in need in the developing world.

4.4.2 Light and Clean Water for Community

According to the World Health Organization, more than 3.5 million people each year die from diseases related to unsafe water and 99% of these deaths occur in the developing countries. Half of these deaths occur to children under the age of 5. The report also says that 40% of the people in the world, especially those in Africa and Asia do not have access to clean drinking water. Clean and safe water is essential to healthy living.

New Vision has developed a water filtering system which accompanies a portable solar light and power pack. It consists of a water filtration pouch inside the box for families to carry water to their homes. The pouch attaches to a water filter and removes 99% of all bacteria to provide clean drinking water. The water filter is created from a technology derived from kidney dialysis. The details are given below.

4.4.2.1 Small Filtration System for Disaster and Remote Communities

- Solar panel: 10W polycrystalline panel with 25 year lifespan
- Battery pack: 12V/9.8Ah lithium polymer with 6–7 years lifespan depending on usage and care
- Water filtration: 5 Gallon (20L) pouch and single water filter
- Light sources: Three Ultra lightweight super bright LED light strips which provide room lighting with 30 year lifespan

The unit has one USB port and costs US$195. Part of the cost is used to send the units to the people in need in the developing world.

4.4.2.2 Large Water Filtration System for Disasters and Remote Communities

- Solar panel: Eight 10W solar panels
- Battery: Eight lithium polymers. It will charge solar LED lights and cell phones with proper USB cable. Battery will not overcharge or discharge and has 6–8 years lifespan depending on usage and care
- Water filtration: 30 Gallons water and storage tank shipping container
- Light sources: Eight solar LED lights

A large water filtration system contains one DC pump which runs off with 60V DC derived from solar panels. It pumps up to 3 gallons of water per minute or 2000 gallons of water per day and has 17 feet of water lift potential. The unit contains six water filters with filter cleaners. One water distribution stand is used which holds the pump stable and provides a distribution system to filter the water. The unit costs US$3000, with optional cart costing US$500. Part of the cost is used to send the units to the people in need in the developing world.

4.5 CASE STUDY: WE CARE SOLAR SUITCASE (YELLOW) FOR MEDICAL RELIEF IN REMOTE AREAS

According to the World Health Organization, 800 women die each day from pregnancy or childbirth-related complications around the world; 99% of these occur in the developing countries. In 2013, 289,000 women died during and after pregnancy, and childbirth in remote areas where the health clinics do not have reliable electricity and reliable lighting. In many developing countries of the world, hospitals and clinics use kerosene lamps and flashlights due to the absence of electric grid or unreliable electricity. This leads to inappropriate medical practices and an increase in pollution. Lack of electricity or irregular electricity affects hospital and communication equipment, which impairs the operation of surgical and delivery wards. This compromises the ability of health workers to provide adequate, safe and timely medical care.

Stachel, an OB-GYN from California, on her visit to Northern Nigeria in 2008 was deeply distressed to observe the unreliable source of electricity during childbirth. The lights in the medical clinics would go out and the physicians would use flashlights to complete the procedure. In 2009, Stachel and Aronson of WCS (We Care Solar) developed a portable solar suitcase shown in Fig. 4.14A that provides power to critical medical lighting, medical devices, laptop computers, and mobile communication devices in remote areas with no access to electricity. The solar suitcase was developed in response to lowering maternal mortality in Nigerian state hospitals as shown in Fig. 4.14B.

After the innovative use of the WCS solar suitcase in the northern Nigerian maternal health clinics, requests for its usage were received from health clinics

(A)   (B)

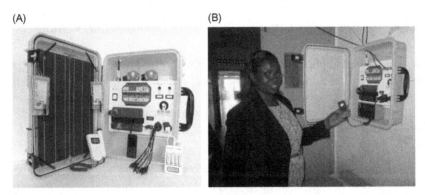

FIGURE 4.14 (A) We Care Solar suitcase for medical; (B) We Care Solar suitcase in a Nigerian Maternity ward. *We Care Solar Suitcase*®.

and health workers from around the world including medical relief work for the 2010 Haiti earthquake. With increasing demand of solar suitcases, various organizations such as The MacArthur Foundation and The Blum Center for Developing Economies supported WCS's programs in education, technology research and northern Nigeria field studies. More than 20 solar suitcases were supplied to the Liberian Institute for Biomedical Research Health for use in primary health care clinics who chose to partner with the World Health Organization. The solar suitcase are currently assembled by a local manufacturer and more than 900 have been sent to 25 countries for medical and humanitarian usage including Mexico, Uganda, Malawi, Sierra Leone, Nicaragua, South Sudan, and recent disasters in Haiti, Bali, and Indonesia. The program will be expanding to regional programs in 2015 to Ethiopia, Tanzania, and the Philippines.

The Solar suitcase, an award-winning system, consists of 40–65 W solar panels, sealed lead–acid battery, inverter, high efficiency LED medical task lighting, a universal cell phone charger, and a battery charger for AA, AAA batteries and provides electricity to 12 V DC devices. The smallest system includes a 40–65 W PV solar panel. A large system with peak available power to 240 W is also available. It produces 1 kW of power during a day of full sunlight. The maternity kit comes with a fetal Doppler. An expansion kit with larger batteries is also available. Its application can range from providing lighting to emergency obstetric care to charging and operating cell phones, lap top computers and medical devices in a remote or disaster ridden area. The factory assembled suitcases cost US$1500 and the US$1200 kit are often purchased by international charitable organizations, NGOs and governments. The commercial system comes in a watertight, protective hard case.

4.5.1 Case Study: We Share Solar Suitcase (Blue) for Lighting in Schools and Orphanages

According to the United Nations Development Program, it is estimated that nearly 1.4 billion people worldwide do not have access to electricity, in addition to the 1 billion that have unreliable electricity networks. This situation has led many NGOs and social entrepreneurs to develop and find products to provide electricity for people living in remote areas with no access to electricity. After the successful launch of the WCS program, a new program called We Share Solar (WSS) was started in 2013 by Aronson, cofounder of WCS, to address educational needs of solar power systems for nonmedical application such as in schools and orphanages. The objectives of the program include using US students to build solar portable system for schools and orphanages in the developing countries. The program also helps US students learn about solar energy and energy efficiency as well a sense of accomplishment by volunteering their time. Aronson of WCS and Alan Jensen of Central Coast High School in Monterey, California, created the curriculum for WSS program.

The WSS suitcase is a complete solar electric system consisting of 20 W solar panels, a battery, a charge controller, switches, and wiring. It can be easily transported and used to illuminate a classroom with only 15 W of electricity, and has the capacity of increasing it up to 200 W of solar power. The WSS suitcases generate electricity by using a solar panel, a 12 V battery and a charge controller. The power produced by each suitcase is enough to light 4–8 lightbulbs and charge a cell phone or a laptop. In this project, funds to buy the solar suitcase kit are first raised by the US students, teachers or school. The students than learn about the workings of circuits and electricity before assembling the suitcase. The students work as a team and select the country, the school, an NGO or orphanage to send the suitcase. This selection process also helps the students to learn about other countries, their standard of living, culture and availability of energy and usage. The students in United States are encouraged to stay in contact with the recipients and document the usage and progress. Training is provided by the WSS partner, after the country and recipients are selected. In order for the recipient to appreciate the solar suitcase, and take care of it, part of the cost is expected from them in some form. Some schools are using medical solar suitcases as a teaching tool in the classroom. More than 45 blue solar suitcases have already been built by US students (shown in Fig. 4.15) as a class project or in a workshops in middle school, high school or community colleges since the program started in 2013.

These suitcases are built to provide light and electricity to charge cell phones, perform emergency medical procedures in the countries without access to electrical grid or having too many power outages. The We Share Solar Program have worked with different schools throughout the United States for different countries worldwide. The details of some of these are given below:

- *We Care Solar, Sacramento county office of Education, Cosumnes River College, Green Technology and Los Rios Community college, California.*

  In 2014, students from seven high schools of Sacramento County Office of Education campuses, Martin Luther King Jr. Technology Academy, Sacramento Job Corps and Visions in Education K-12 Charter School,

FIGURE 4.15 US students who assembled WSS suitcase for schools and orphanages in the developing countries. *Reprinted with permission. © 2015 Home Power Inc., www.homepower.com.*

California participated in two workshops to build portable lifesaving solar lighting systems for developing countries. The goal of the workshops was to extend building portable solar systems both to act as an educational tool for green energy as well as a much-needed energy source in developing countries. The workshop was held at Cosumnes River College and students assembled 35 portable solar suitcases which were delivered to schools and medical clinics in Uganda and Haiti that was hit by severe earthquake in 2010. The suitcases were sent to developing counties by the team of Green Technology solar suitcase project. The workshops were sponsored by Cosumnes River College and Sacramento County Office of Education in partnership with We Care Solar and Los Rios Community College District.

- *We Share Solar, Sacramento State University, Cosumnes River College, the Northstate Building Industry Association, SMUD, Green Technology and the Sacramento County Office of Education, California.*

In 2013, approximately 40 local students from Sacramento County, California, participated in a green technology workshop to build solar suitcases for life saving energy in developing countries. During the workshop, participants ranging in ages of 15–25 from Gerber Jr. /Sr. high school and the Boys and Girls Club Links received hands-on training in green energy technology and also learned about social consciousness. The workshop was organized in partnership with We Share Solar, Sacramento State University, Cosumnes River College, the Northstate Building Industry Association, SMUD and the Sacramento County Office of Education. Green Technology, an NGO, sent 20 of the assembled suitcases to provide portable lights to impoverished countries of the world. Green Technology trains students in the use of renewable resources by focusing and developing student's skills in environmental protection, clean energy, energy efficiency and also provides career opportunities for youth from underrepresented communities.

- *Valencia College, Florida and Valencia Foundation.*

In 2014, students shown in Fig. 4.16A at Valencia college, Florida, a 2 year community college, from Dr. Hall's class on "Introduction to Alternative and Renewable Energy" assembled four suitcases. The cost was covered by a grant of $5000 from Valencia Foundation. Three suitcases were sent to NGOs in Haiti, Guatemala and Costa Rica, and the fourth destined for Kenya was delayed due to ongoing conflict in the country. The suitcase for Haiti was used in an orphanage school to provide emergency power for charging cell phones, laptops and rechargeable batteries for headlamp and flashlights by children and teachers. Two engineering students from the class decided to send one suitcase to Antigua, Guatemala, through an NGO called the Integral Heart Foundation. Their decision was based on the fact that in rural Guatemala, an average family spends US$14–20 per month on candles and US$20 to run two incandescent light bulbs. As a result solar is a good alternative and it will be used to provide light for children to finish their homework (shown in Fig. 4.16B) at night without the risk of burning

(A)   (B)

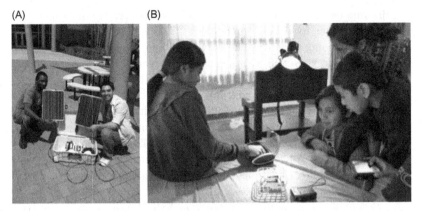

FIGURE 4.16   (A) Students of Valencia College; (B) Light for the Children of Guatemala.

candles, helping to conduct medical procedures at night, and saving family money from buying inadequate candles for lighting.

The third suitcase was sent to Rancho Mastatal in Costa Rica, which is run by Peace Corps veterans. The suitcase was used to demonstrate the use of solar power and how it can be employed in sustainable living to the staff of Rancho Mastatal which is an environmental learning and sustainable living center. It is a rural sanctuary and lodge located on the edge of the last remaining virgin rainforest of Costa Rica's Puriscal County. The mission of the center is to practise, promote and teach about living responsibility in the tropics and the implications of disappearing tropical forests. The center also helps in conservation, community growth, education, natural medicine, food security, and production.

The last and fourth suitcase was delivered to the orphanage school in Thika, Kenya, on Jul. 2014. This was achieved by a member of Northland: A Church Distributed, and their team, who volunteered to take the solar suitcase on their upcoming trip to Kenya. It was built by three students who hoped that the suitcase will serve as a token of appreciation and respect for helping the children in conflict by people from thousand of miles away. The Northland: A Church Distributed is a nontraditional church located in Orlando, Florida.

- *We Care Solar and Technical Museum of Innovation, San Jose, California.*

We Care Solar launched the We Share Solar program to build ten solar suitcases by 40 students in collaboration with the Technical Museum of Innovation in San Jose, California. The day long workshop was supported by Applied Materials and instructed by 12 We Share Solar program teachers. It was attended by local girl scouts, Third Street Community Center and teams from the Challenge program of the museum. The workshop was held at the newly-opened TechLab of the museum. Before assembling the suitcases, the students learned about the workings of electrical circuits, energy poverty, and sustainable development. Upon the completion of suitcases the students working in teams commissioned the system by testing various components including LED lights, system charge controller, and by charging their own cell phones.

The assembled suitcases were sent to the orphanage schools in Uganda and Sierra Leone that do not have light. The US students wrote notes for these students who do not have light to study and finish their home work after sunset. Some of these suitcases will be installed in a girl's dormitory at New Hope Orphanage in Bugiri, Uganda, in order to provide a safer environment for girls to study and play after the sunset. Other suitcases will be sent to Sierra Leone for illuminating the library of the Children in Crisis Primary school in Upper Allentown. With the first light for a school with over 470 students, the students will be able to study at night and the teachers will be able to hold workshops and community meeting in the evening. The whole workshop provided an excellent education experience for US students and provided light to the youth in a country without access to electric grid.

- *Middle Schools and Pace Center for Civic Engagement at University of Princeton, New Jersey.*

Four teams of middle-school students assembled four solar suitcases for the We Share Solar program as part of the Community House STEM Summer Camp at Princeton University, NJ. This project was completed in collaboration with the Pace Center for Civic Engagement, and was supported by two Princeton alumnis: Les Gutierrez'84 and University Trustee Jaime I. Ayala'84. The assembled suitcases with lights were installed by Dallas Nan'016 in an elementary school at Apia, Philippines as part of a summer internship with Stiftung-Solarenergie, a German NGO. Stiftung Solarenergie-solar energy Foundation Philippines is a nonprofit organization dedicated to fighting poverty in the Philippines by proving off-grid villages with sustainable access to solar energy. Their efforts give users the light and hope necessary to break the cycle of poverty, and paving the way for further development. The Director of the Pace Center for Civic Engagement, Kimberley de los Santos, was pleased with the pathways for US STEM students to explore relationship and forming connection between middle-school students, Princeton university alumni and undergraduate students at Philippines.

- *We Share Solar, Princeton Day School and The Stuart School, Princeton University PACE Center, New Jersey.*

The codirector of We Share Solar, Gigi Goldman, organized a workshop with the help of Linda Gaffney and Warcheerah Kilima, in snowy weather on Feb. 2014 at Princeton, NJ. The purpose of the workshop was to train volunteers teams headed to Peru and Haiti to practise techniques utilizing the following new constructions:

1. A wall for solar suitcase installations
2. A knee-high corrugated tin flat to practise drilling and panel layout
3. An eight-foot-high hut with corrugated tin roof

Training also involved the practice of safety skills, use of power tools and knowing the skill of mounting solar suitcases to wooden walls and safely attaching them to rooftops. Two teams of students who participated in the workshop and built the solar suitcases belonged to local Princeton Day School and The Stuart School as a part of We Share Solar's student education program.

After the training and building the solar suitcase, the first team traveled to Peru in Mar. 2014 to install three solar suitcases in the community centers of the remote areas of Abra Malaga, Corpani Penas. The second team traveled to Haiti in Mar. and Jun. 2014 also installed solar suitcases in the community centers. The second team was accompanied by members of Princeton University PACE Center and Princeton Institute which is an educational nonprofit organization focused on building partnership between Princeton and Fond Parisien, Haiti.

BIBLIOGRAPHY

[1] L.S. Beaty, Valencia Students Create Portable Solar Generators for Needy Villages. <http://news.valenciacollege.edu/valencia-today/valencia-students-create-portable-solar-generators-for-needy-villages/>, 2014.
[2] Bring Your Creative Project to Life, Kickstarter. <https://www.kickstarter.com/>, 2014.
[3] P. Danko, Solar Charger with Tracking Comes in a Briefcase. <http://earthtechling.com/2013/08/solar-charger-with-tracking-comes-in-a-suitcase>, 2013.
[4] K. Davidson, Energizing Education. Home Power, Issue#156. <http://www.homepower.com/articles/solar-electricity/design-installation/energizing-education>, 2013.
[5] G. Goldman, We Share Solar Launches at the Tech! <http://wecaresolar.org/we-share-solar-launches-at-the-tech-2/>, 2013.
[6] M.C. Go, Light Bearers to the Father's Houses in Uganda, Africa. <https://goproject.org/2014/08/power-and-light-bearers-to-the-fathers-houses-in-uganda/>, 2014.
[7] B. Harris, Solar Suitcase Project' Uses Solar Technology to Impact Local Students and the World. <http://sacramentopress.com/2014/04/02/solar-suitcase-project-uses-solar-technology-to-impact-local-students-and-the-world/>, 2014.
[8] High Efficiency Foldable Solar Panel-75 Watt. <http://www.powerenz.com/store/index.php?_a=viewProd&productId=269>.
[9] J. Keller, U-Mass to Help Army Develop Electricity-Generating Fabric for Tents and Backpacks. <http://www.militaryaerospace.com/articles/2013/03/UMass-photovoltaic-fabric.html>, 2013.
[10] Lighthouse 250 Lantern, <http://www.goalzero.com/p/180/lighthouse-250-lantern>, 2014.
[11] Light and Water, <http://www.nvre.org/pdf/light-and-water.pdf>, 2014.
[12] Light & Power, <http://www.nvre.org/pdf/light-and-power.pdf>, 2014.
[13] LINORTREK: OTG-15W— Portable & Rechargeable Solar Power Generator for Emergency Survival. <http://www.linortek.com/otg15w-emergency-solar-power-generator/>.

[14] Maternal Mortality, Fact Sheet No 348. World Health Organization. <http://www.who.int/mediacentre/factsheets/fs348/en/>, 2014.

[15] Meeting the MDG (Millennium Development Goal) Drinking, Water and Sanitation, the Urban Rural Challenge of The Decade. World Health Organization and UNICEF. <http://www.who.int/water_sanitation_health/monitoring/jmpfinal.pdf>, 2006.

[16] R. Miller, Quality of Life Improved by Solar Lanterns: New Survey of Rural Haitians Documents Multiple Benefits of Solar Lights. GlobeScan <http://www.globescan.com/>, 2014.

[17] T. O'Hara, R. Nunes, Rancho Mastatal: Environmental Learning Center and Lodge. <http://www.ranchomastatal.com/pages/links/page.php?Grouping=Home&PageName=home>, 2014.

[18] Portable Solar Power Pack & Light. <http://www.nvre.org/powerandlight/index.htm>.

[19] Portable Systems for Remote Locations and Emergency Disaster Response. Sunshine Works. <http://www.sunshineworks.com/solar-water-purification-systems.htm>.

[20] POWERENZ LFP40 Waterproof System. <http://www.powerenz.com/store/index.php?_a=viewProd&productId=284>.

[21] M. Quinn, D. Prieto, Integral Heart Foundation. <http://www.integralheartfoundation.org>, 2014.

[22] Ray of Life, <http://www.nvre.org/pdf/ray-of-life.pdf>, 2014.

[23] R. Seaman, New Vision Renewable Energy (NVRE). <http://www.nvre.org/>, 2014.

[24] L. Staachel, We Share Solar Winter Training Preps Youth for Solar Suitcase Installs in Warmer Climates, 2014.

[25] J.T. Smith, Tested: Goal Zero Lighthouse 250 Lighting the Way, From Yellowstone to Haiti. <http://gearpatrol.com/2014/04/16/tested-goal-zero-lighthouse-250/>, 2014.

[26] SolarWrap 250, <http://www.outdoorlife.com/blogs/hunting/2013/06/gear-test-solar-panels>, 2014.

[27] Sacramento Office of Education County, Local Students Build Solar Lighting Systems to Benefit Developing Countries: Portable Cases to Be Used by Schools and Medical Clinics in Haiti and Uganda, Sacramento, CA. <http://www.scoe.net/news/library/2014/april/01solar_cases.html>, 2014.

[28] Solar Suitcase Project, Green Technology. <www.greentechedu.org>.

[29] Solar Suitcase for Emergency/Survival. <http://inhabitat.com/super-portable-solar-briefcase-lets-you-carry-solar-power-everywhere-you-go/>, 2014.

[30] SCOE Students "Plug In" to Solar Case Project: Environmental Initiative Benefits Groups in Uganda and Haiti. <http://www.scoe.net/news/library/2013/march/25solar_cases.html>, 2013.

[31] Solar Water Purification Portable Systems for Remote Locations and Emergency Disaster Response: Sunshine Works. <http://www.sunshineworks.com/solar-water-purification-systems.htm>.

[32] Sunshine Solar Briefcase Charger Amorphous 13 Watt, Sunshine Solar. The Sun Belong to Everyone. <http://www.sunshinesolar.co.uk/khxc/gbu0-prodshow/SS936A.html>.

[33] UN Water World Day 2013: International Year of Water Cooperation. Facts and Figures. Coordinated by UNESCO. <http://www.unwater.org/water-cooperation-2013/water-cooperation/facts-and-figures/en>.

[34] Ultralight Portable Solar Charger Panel and Power Module. <http://www.gofastandlight.com/Ultralight-Portable-Solar-Charger-Panel-and-Power-Module/productinfo/PO-R80005/>, 2014.

[35] We Care Solar. <www.wecaresolar.org>.

[36] G. White, Billion Daily Affected by Water Crisis. Water.org. <http://water.org/water-crisis/one-billion-affected/>.

[37] What We Do: Achieving Universal Energy Access. United Nations Foundation. <http://www.unfoundation.org/what-we-do/issues/energy-and-climate/clean-energy-development.html>.

[38] J. Yago, Portable Power, Issue #134. <www.pvforyou.com>, 2012.

[39] 11 Facts About Water Around the World. <https://www.dosomething.org/facts/11-facts-about-water-developing-world>, 2014.

[40] 30 Watt Military Grade Solar Panel. <http://www.endlesssunsolar.com/products/30-watt-military-grade-thin-film-folding-solar-panel>.

Chapter 5

Fixed Standalone PV Systems for Disaster Relief and Remote Areas

5.1 INTRODUCTION

Fixed standalone PV systems are essentially immobile systems and once installed, cannot be easily moved on wheels, unlike mobile PV systems, or carried manually as in the case of portable standalone PV systems. Fixed standalone PV systems have been deployed to provide electricity to power radio stations, health clinics, shelters, and homes at disaster sites before utility electricity is restored. Such systems can also be used for more than 1.3 billion people, with the majority of them living in remote areas with no access to electricity. Fixed standalone PV systems differ in size, capacity, and applications from mobile and portable PV systems.

A fixed standalone PV system converts the sun's energy into electricity by using PV cells that are designed and sized to supply certain DC and/or AC electrical loads. The simplest type of standalone PV system is a direct-coupled system where the output of PV array or module is connected directly to DC load. Since there is no electrical energy storage (batteries) in direct-coupled systems, the load only operates during sunlight hours. This makes these designs suitable for common applications such as ventilation fans, water pumps, and small circulation pumps for solar thermal water heating systems. In order to store energy for the night or cloudy days, a standalone system with storage is used. DC power that is produced in a fixed standalone PV system is stored in deep cycle batteries and converted into AC by an inverter. In some cases, where it is important that power is always available, some standalone systems, known as PV-hybrid systems or island systems, may also have another source of power such as a wind turbine, the use of biofuel or a diesel generator.

Fixed standalone systems can provide electricity required for emergency relief. In addition, they can also provide electricity for water pumping, drinking water and livestock water pumping, water purification, irrigation pumping, village electrification, refrigeration of medical supplies and food, power for communication equipment, base stations, street lighting, and others. This chapter

Standalone Photovoltaic (PV) Systems for Disaster Relief and Remote Areas.
DOI: http://dx.doi.org/10.1016/B978-0-12-803022-6.00005-8

discusses the use of fixed standalone system in some of the applications given below:

- Solar-powered water pumping systems
- Design solar-powered pumps
- Application of solar-powered pumps for:
  - supply of water to homes and villages
  - irrigation (drip irrigation system)
  - livestock watering
- Water purification systems:
  - Ultra-violet (U-V) sterilization
  - Solar-powered reverse osmosis water purification systems
  - Solar-powered Ultrafiltration system
- Direct solar treatment for purification of water:
  - Solar water disinfection (SODIS)
  - Solar pasteurization
  - Solar water distillation using solar Stills
- Case study of standalone PV system for mobile communication in remote areas:
  - Solar-powered base stations and their benefits
  - Examples of solar-powered base stations for disaster relief and remote areas
  - Solar-powered base stations in India

5.2 SOLAR-POWERED WATER PUMPING SYSTEMS

When disaster strike, the normal supply of drinking water is shut down or becomes contaminated. A solar-based electrical pump can be used to extract water from underground, a stream, or river which is transported to a water tank ready for the local community to drink. In remote areas and the developing world, solar pumps are used where there is no electricity or water sources are scattered such as cattle ranches or village water systems. According to the United Nations' World Water Development report, more than a billion people worldwide do not have access to safe drinking water. A solar pump is used to fill a central water storage or a tank that is located at a high point of the property. The stored water can then be distributed by gravity feed to a network of pipes to individual stock tanks for irrigation, drinking or livestock watering. Water pumping systems can also be oversized to store enough water which could be used for emergency during drought.

A solar-powered water pumping system consists of solar panel which produces DC electricity when exposed to sunlight. The DC current is collected at the output of the solar panel and is applied to a DC pump for pumping water. There are two basic types of solar-powered water pumping systems: direct-coupled and battery-coupled.

5.2.1   Direct-Coupled Standalone Solar-Powered Water Pumping Systems

A direct-coupled solar-powered water pumping system consist of PV panels, pump controller, pressure switch, DC water pump and storage as shown in Fig. 5.1. The direct-coupled solar pumping system is fairly easy to install in which water will be pumped when there is sunlight. In the absence of using pumped water, a tank is filled for later use.

The electricity from the PV modules is sent to the pump through a controller, which in turn pumps water through a pipe to where it is needed. The amount of water pumped is dependent on the amount of sunlight hitting the PV panels and the type of pump used in the system. The amount of water pumped changes throughout the day since the intensity of the sun and the angle at which it strikes the PV panel is different throughout the day. This affects the efficiency of the pump. The pump operates at 100% efficiency from late morning to late afternoon on bright sunny days with maximum water flow. However, the PV panels will produce 25% lower efficiency during early morning and early afternoon causing minimum water flow. The pump efficiency will drop off even more during cloudy days. This variation in efficiency throughout the day may produce a fairly constant voltage but current output changes dramatically with changes in light intensity. The pump motor can run on a lower voltage but it needs certain amount of current to start. To compensate for these variations and to obtain a match between the pump and PV modules, a pump controller is used which allows the pump to start and run at reduced output in weak-sunlight periods. A pressure tank and pressure switch or a float switch is generally used to reduce cycling on and off of the pump. Depending on the type and size of solar-powered water pumping system the function of pump controllers can range from not using any controller to sophisticated smart controllers for the following tasks:

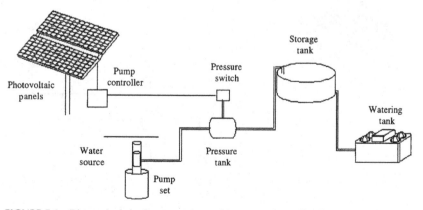

FIGURE 5.1   Direct-coupled solar-powered pumping system. *Agricultural Extension Services of the University of Tennessee.*

- Improving pumping performance by adjusting voltage and current at lower solar radiation levels
- Limiting power to diaphragm pump motor in order to keep it from being damaged
- Providing manual disconnect switch between PV modules and pump motor
- Allowing automatic disconnect by having a pressure tank and pressure switch between PV modules and pump motor when storage tank is full.

5.2.2 Battery-Coupled Standalone Solar-Powered Water Pumping System

In a battery-coupled solar water pumping system shown in Fig. 5.2, batteries are used to extend the pumping over a longer period of time by providing a steady operating voltage to the DC motor of the pump.

As a result, the system can still deliver a constant source of water during the night and low light periods. The primary function of a pump controller in a battery-coupled pumping system is to boost the voltage of the battery's bank to match the desired input voltage of the pump. In the absence of a pump controller, the PV modules' operating voltage is dictated by the battery bank and is reduced considerably from levels which are achieved by operating the pump directly. A pump with an optimum operating voltage of 30V would pump more water tied directly to the PV panels than if connected to the batteries. In this case the use of a pump controller with a 24-V input would step the voltage up to 30V, which would increase the amount of water pumped by the system. Battery-coupled PV pumping systems have the following drawbacks compared to direct-coupled pumping systems:

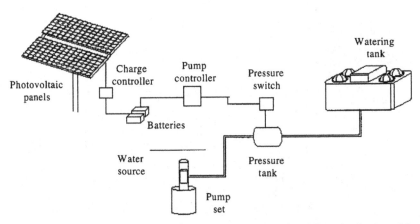

FIGURE 5.2 Battery-coupled solar water pumping system. *Agricultural Extension Services of the University of Tennessee.*

- A battery-coupled PV system is more complicated because of additional components.
- Requires additional maintenance and is more expensive.
- Addition of batteries reduces the overall efficiency of the system as the operating voltage is dictated by the batteries and not the PV panels.
- Voltage supplied by the batteries can be 1–4 V lower than the voltage produced by the panels during maximum sunlight conditions depending on their temperature and how well the batteries are charged.

Although the reduced efficiency can be minimized with the use of an appropriate pump controller, most of the system in usage is direct coupled.

5.3 DESIGN OF A STANDALONE SOLAR-POWERED PUMPING SYSTEM IN EIGHT STEPS

The design of solar-powered pumping systems as shown in Fig. 5.3 requires determining the size of the pump and PV system used to power the pump.

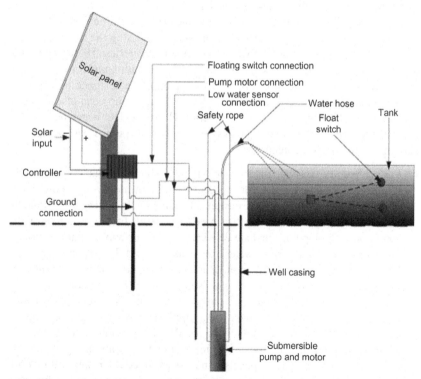

FIGURE 5.3   Schematic diagram of a typical solar-powered pumping system. *"Solar Photovoltaic Water Pumping for Remote Locations," University of Wyoming, 2006. Reprinted from K. Meah, S. Fletcher, S. Ula, Solar photovoltaic water pumping for remote locations, Renew. Sustain. Energy Rev. 12 (2) (2008) 472–487, with permission from Elsevier.*

The size and type of the pump will depend on several factors including water supply, water need, pumping head, and others. The PV panel depends on the voltage needed to run the pump which will then determine the number of panels. Steps to designing such a system are:

1. Determine whether the pump is surface-mounted or submersible
2. Determine the required water flow
3. Determine the pressure needed to transport the water from source to destination
4. Select a pump that will provide the needed flow and pressure
5. Select the right size pipe for the pump
6. Determine power needs for the desired pump and size of PV systems
7. Select the correct solar array mounting
8. Using water storage and water level sensors

5.3.1 Step 1: Determine Whether the Pump Is Surface or Submersible

Solar pumps use DC current from PV panels and/or batteries to move water from one location to the other. Most solar-powered water pumps are low volume, pumping 2–4 gallons of water per minute. This keeps the cost of the system low by using a fewer number of solar panels and using the entire daylight period to pump larger amount of water or charge the batteries. These pumps are designed to work effectively without stalling or overheating under low sunlight and reduced voltage. The most efficient solar pumps are designed to use solar power which operates between 12 and 36 V DC. AC pumps can also be used by converting DC into AC using an inverter. However, DC water pumps use one-third to one-half of the energy of conventional AC pumps. DC pumps are classified in two major categories:

- Surface pumps: these are displacement pumps which use diaphragms, vanes, or pistons to seal water in a chamber and force it upward through a discharge outlet. Displacement pumps are designed to maintain the lift capacity throughout the solar day at slow, varying speeds that result from varying sun light conditions. These pumps include delivery, pressure or booster pump. Surface pumps are often used to draw water from a lake or pond. These pumps are used for moving water through the pipeline and are located at or near water surface.
- Submersible pumps: are centrifugal pumps which use a rotating impeller that adds energy to the water and pushes it into the discharge outlet. These pumps start gradually and their flow output increases with the amount of current, which can be tied directly to the PV panels without connecting to a battery or controls. In order to achieve efficient operation, a good match between the pump and PV panels is necessary since their output drops off at reduced speed. Centrifugal pumps are used for low lifts, often for well pumping. The submersible pumps are placed down a well and are not subject

to extreme weather conditions. Its location makes it more reliable and does not need special protection. Some surface pumps can develop high heads and are suitable for moving water long distances or to high elevations.

5.3.2   Step 2: Determine the Required Water Flow

Flow rate or water demand is the amount of water needed to irrigate, supply to livestock, domestic and/or village use. It refers to the maximum amount of water a solar pump can flow at a given amount of time and is usually measured in gallons per minute (GPM). The maximum flow rate is defined by assuming that the water is being moved horizontally without resistance. The flow rate will be decreased if a pump has to pump through a long pipe, raise the water from below or provide pressure for other reason. In other words, the amount of water pumped over a given period of time decreases as the vertical pumping distance increases.

Daily water needs for livestock varies with the animals' ages and size, location, distance to water, air temperature and others. Water demand for domestic household use is variable depending on climate and usages but it is typically around 75 gallons per person in United States for drinking, cooking, and bathing. Irrigation water demand depends on crops, season, delivery method, and evaporation. Agriculture watering is usually greater in summer seasons when solar has its highest capacities.

Typical water requirements are:

- One Person for all purposes: 50–75 Gallons per day (GPD)
- One Milking Cow: 35 GPD
- One Cow/Calf Pair: 35–40 GPD
- One Horse or Dry Cow: 10–20 GPD
- One Sheep: 2 GPD
- One Hog: 4 GPD
- 100 Chickens: 4 GPD

After calculating the daily water requirement, determine the number of hours of bright sunlight which powers the pump. This can be done by locating the nearest latitude coordinate of the water source, the flow rate is given by the following formula.

$$\text{flow\_rate (in GPM (gallon per minute))} = \frac{\text{daily\_water\_needs (gallons)}}{\text{hours\_sunlight\_per\_day} \times 60 \text{ minutes per hour}}$$

5.3.3   Step 3: Pressure Needed to Move Water From Source to Destination

For purposes of designing a solar pumping system, the total dynamic head (TDH) is defined as the effective pressure the pump must operate against and can be thought of as the work that the pump must overcome to move a certain amount of water. It is most often expressed in either feet per head or psi (pounds per square inch) and is determined by the total vertical pumping distance between the water source and the water tank. For some residential applications, this distance can be more than 100 feet, while for agricultural application, this distance is between 10 and 20 feet. To pump water from underground, knowing a pump's maximum pressure allows calculating the deepest water that it can pump to the surface. System pressure is roughly equal to an increase of 1 psi for every 2.31 feet of elevation head. In other words, PSI can be converted into feet per head by dividing it by 2.31. TDH is expressed in feet and is the sum of the following four factors:

- Pumping level: This is the level of required water flow in the well and is measured in feet. The static water level is the distance from the top of the well to the surface of the water level. Drawdown is the drop in water level which results from pumping water from the well.
- Vertical Rise: This is also referred to as elevation; it is the number of feet of vertical rise from the well to the ultimate delivery point such as storage tank. It could be a small rise to the top of a holding tank or a large rise to the top of a hill.
- Friction Losses: It is the loss of pressure due to flow of water in pipe measured in feet from the pump to the point of use. These losses are caused due to many factors such as type of pipe, length and diameter of the pipe as well as type of pipe fitting used. Friction losses also depend on the flow rate and are greater at the higher rate of flow. Tables are provided to determine the friction losses.
- Tank Pressure. It is the operating pressure of the storage tank and is measured in feet. Large storage tanks used in solar pumping systems have zero tank pressure as they are not pressurized. However the systems running with battery power can be used to pump into a pressurized storage tank.

$$\text{Total Dynamic Head} = \text{Pumping Level} + \text{Vertical Rise} \\ + \text{Friction Loss} + \text{Tank Pressure}$$

5.3.4 Step 4: Determine the Right Size of Pipe

Most PV applications will be pumping water at low flow rates of 1–5 GPM and these flow rates will not have sufficient water velocity through a large pipe to keep suspended solids from settling out into the bottom of the pipe. Therefore, a smaller pipe measuring ½ inch to 2 inches is typically sufficient for PV pump without much friction losses which makes it cheaper and more efficient. To overcome the pressure loss due to friction within the pipe, a larger diameter distribution pipe can be used. Typically, the largest diameter pipe is used to get water from the pump to the destination, then smaller branches of pipe can be used at the final stage of water exiting the system through drip taps or drip emitters in an irrigation system.

5.3.5   Step 5: Select a Pump That Will Provide the Required Flow and Pressure

Having found the flow rate and TBH, the pump can be chosen from specifications given in the form of tables and graphs available online or in the product literature provided by the manufacturers. Pumps are rated to produce a certain flow at a certain pressure when supplied a certain amount of power. To choose a pump, the first step is to refer to pump specifications that rate the pump output in terms of pressure and flow. It is important to make sure during referencing that the pump will provide the amount of pressure which is determined to be needed. The next step is to move along the pump curve graph to determine whether or not the pump will produce the required flow at the needed pressure. This will give you the pump type and the model. During referencing, if the flow is not sufficient at the particular pressure needed, then a different model or different type of pump should be used. If the system is operated at various flow rates, a system curve can be used. These pumps are low volume pumping at an average of 2–5 gallons and are either kinetic/centrifugal or positive-displacement. Many of these pumps are capable of meeting the requirements of pumping from a surface source or a well. The commonly used pump for a well application is a DC submersible pump with a voltage range of 15 V to more than 45 V (higher voltages are used for deep wells), 3–5 A (1/4 to 2 HP) range. In order to maintain the lift capacity of the pump throughout the solar day at varying speeds resulting from changing light conditions, the majority of the pumps manufactured are positive-displacement pumps.

5.3.6   Step 6: Determine Power Needs for the Desired Pump and Size of PV Systems

The next step is to determine the voltage (12, 24, and 36 V) and power in Watts required by the selected pump to pump the water. An extra 25% power is also required to aid in kick-starting the pump which is critical for installations due to commonly occurring overcast skies.

PV arrays are used to power the pump by wiring individual PV panels in series or in parallel to obtain the required voltage or current needed to run the pump. The individual solar panels come in all sizes ranging from 50 to 80 W with the resulting array wattage ranging from 100 to 320 W. For example, a PV module rated at 50 W operating at voltage up to 17.4 V and a maximum current of 3.11 A can be wired in series to increase output voltages and in parallel to increase current while also increasing total power. The maximum voltage output from two of these panels in series is 34.8 V and currents will remain the same as for one module. Thus, a 24-V DC pump requires a minimum of two, 12-Volt panels wired in series. The output current and hence the power will fall at a relative level if the sunlight changes on a cloudy day. Larger PV panel surface areas act as a linear current booster which will allow the pump to turn on earlier and later in the day and also in relatively low light conditions. In direct-coupled systems where the panel is directly wired to a pump, 20% more wattage than specified is usually required. Most of the pumps need two to four PV modules using fixed structure or a tracking system to maximize power by enabling the PV array to follow the sun movement. However, most systems currently use fixed structures (pole mounting) facing south. The advantages of both mounting is discussed in the next section.

5.3.7 Step 7: Selecting the Correct Solar Array Mounting

There are two ways to mount solar panels as a standalone system:

- Fixed Structures
- Tracking System

A fixed structure is installed facing true south at a certain tilt to receive maximum light throughout the year. Fixed mounts for installing solar panels are less expensive and can tolerate higher winds. In the fixed structure, panels should be oriented to face true south (not magnetic south) and panels facing 30 degrees away from south will lose approximately 10–15% of their power output. Hence, the tilt angle, which is the angle between the plane of the solar panel surface and the ground, needs to be adjusted. In order to collect maximum energy, the panel surface should be perpendicular to the sun or pointing directly toward the sun. One way of selecting the tilt is to adjust it for maximum power in the winter and keep it fixed for the rest of the year. The other way is to adjust the tilt angle to latitude minus 15 degrees for the Summer and latitude plus 15 degrees for the Winter. The tilt can also be set equal to latitude for year-round operation.

In tracking systems, panels follow the sun and increase the power output by keeping the array pointed at the sun throughout the day. This is more expensive than the fixed systems but can substantially improve the amount of power produced by enhancing morning and afternoon performance. The working of a tracking system is based on a stored liquid which is warmed by sunlight and flows through tubing to the opposite side of the tracker. It is the weight of the liquid that causes tracker to the opposite side of the tracker without using any electricity. Other benefits of the tracking system include lower PV required panels, and the reduction in pump stalling due to low light conditions during early morning and late afternoon at low tilt angles. Tracking systems work best in Summer and for centrifugal pumps where water yield drops exponentially with a drop in power. Although tracking systems enhance the power output of the panels by capturing more sun energy, they are more complex and expensive to maintain. It is usually recommended to use more PV panels which may require less maintenance than employing tracking system.

Shading from trees, weeds, and other obstructions can cast shadows on the solar panels and limit power output. It is important to avoid these obstructions when choosing a site for the panels, especially in the Winter when the arc of the sun is lowest over the horizon. It is also important to protect PV panels from damage caused by animals. This can be accomplished by making a fence around the PV module site. It is also desirable to place the solar array as close as possible to the pump to minimize wire size and installation cost.

5.3.8   Step 8: Using Water Storage and Water Level Sensor

All solar water pumping systems use some type of water storage. The purpose of water storage is to store water rather than electricity in batteries, thereby reducing the cost and complexity of the system. The storage capacity should equal 3–10 days of water depending on usage and climate of the region used. In a cloudy climate 10 days of storage may be necessary for domestic usage while for sunny climates 3 days of storage may be sufficient for livestock watering. Five days of storage may be required for irrigating a home garden. The tanks for water storage are often made of food-grade plastic or polyethylene. Other material such as concrete and stainless steel are also used which prevent sunlight penetration for controlling algae growth in the stored water. The tanks are typically placed at a high point on the property or on a hill for gravity feed to drinking troughs, drip irrigation applications or seasonal grazing. Storage tanks shown in Fig. 5.4A located at least 50 feet in elevation above the upper most drinking troughs can achieve a minimum pressure of 20 lbs psi. In order to maintain cooler water temperature, tanks can be buried for year-round use or be placed aboveground in shady wooded areas.

To disconnect the pump when storage tank is full, a float switch is installed inside the tank which controls the pump by sensing the level of water as shown in Fig. 5.4B. This is enabled by running a wire along with the distribution pipe from the switch to the pump controller.

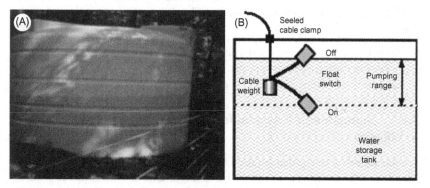

FIGURE 5.4   (A) Water tank for water storage; (B) Float switch inside water tank.

5.4   APPLICATIONS OF STANDALONE SOLAR-POWERED WATER PUMPING SYSTEMS

A solar-powered water pumping system consist of PV panels, a pump, a storage tank and controllers. It is one of easiest solar systems to install as it does not require a battery or a battery charger. The pumping system works when the sun is shining. A storage tank can store the excess water which can continue to supply water during night or when it is cloudy. Solar pumps are specifically designed to accept DC power which uses the absolute minimum of electrical power to pump in cloudy conditions. These pumps are capable of providing water in areas where there is no electricity or water resources are spread over a long distances. Solar pumps in these situations are dependable and cost-effective, especially when transporting fuel is either expensive or difficult. Solar pumps move a small volume of water over an extended time period of time, requiring less power which minimizes the size and cost of PV panels. For smaller pump sizes, the flow rate can be less than 1 GPM which will move continually from 9 am to 3 pm for sunny locations. This low flow rate of 1 GPM provides over 350 GPD of water from most of the wells except the deepest ones. Solar pumps can work from most locations and PV panels can be mounted anywhere but they should face in a southerly direction in North America for maximum sunlight. In livestock watering and farm applications, the PV panels and pump controllers should be mounted on raised platform like a pole to stay above snow drifts and potential damage from animals. The use of pole mounting also makes it easier to adjust the PV panels' tilt and east–west orientation during initial setup to achieve the best overall performance for the whole year. The tilt for most applications is equal to the latitude at which a solar-powered pumping system is located.

Solar-powered pumping systems provide maximum capacity during warm sunny days and can be suitably used for supplying water in remote areas, livestock grazing, and farming operations instead of other forms of alternative energy. These systems are durable, need minimal maintenance, can be mobile, and offer long-term economic benefits. The following applications will be discussed:

- Supply of water to homes and villages
- Irrigation (drip irrigation system)
- Livestock watering

5.4.1   Standalone PV System for Water Supply to Homes or Villages

A solar-powered pump can be used to provide a constant supply of water to homes or villages in the aftermath of disasters and in remote areas with no access to electricity. The difference between the domestic water supply and village water supply is the size of the storage and volume of water. In domestic water supply systems, the storage reservoir is located above the house. As a result, gravity provides the water pressure to the faucets when a tap is opened. Most of the domestic water systems uses a single submersible pump which will supply water to a pressure system located at the house or pump building.

Village water supply systems may require pumping into larger storage tanks where it will then gravity feed to the village. Additional pumps may be used in some systems to pump water into storage tanks. In the case of cloudy days, the village water systems may require back-up generators to pump water as shown in Fig. 5.5. The water for drinking purposes will require filtering and purification. There will also be a need to store water up to 5 days for cloudy days or during the rainy season.

5.4.2   Standalone PV System for Drip Irrigation System

Solar-powered water systems in irrigation can be categorized in two different types. The first type is used for large areas of irrigation where water is flooded into wide areas. This system uses a large quantity of water since some of it is evaporated. The second type is called micro-drip irrigation where water is fed directly to the plants through pipes. This type uses much less water and losses through evaporation are small. Water demand for irrigation varies throughout the year depending on the type of climate and agriculture. During the irrigation seasons, water demand is more than average and it depends on the system of water distribution. Water loss can be directly related to watering of the crops. The water source for irrigation may be a nearby river, stream, pond or a bored well. The pumping system generally use larger centrifugal pumps or shaft driven down whole pumps. For example, trickle/drip irrigation uses much less water compared to the conventional method of irrigation.

An efficient method of distributing water with minimal losses is solar-powered drip irrigation systems for farmers on sunny and hot days. In drip irrigation, the water and fertilizers are applied directly to the root zone of plants by applicators in the form of emitters, porous tubing, or perforated pipes. The applicators which operate under low pressure can be placed either on the

FIGURE 5.5   Solar-powered water supply for homes. *Credit "Jennifer Burney".*

surface or buried underground. Studies have shown that this form of irrigation has resulted in 100% yield gain, 40–80% saving in water and other savings of fertilizers, pesticides, and labor compared to conventional irrigation. In addition to these advantages, drip irrigation is less vulnerable to adverse weather such as a drought. A solar-powered drip irrigation system uses solar-powered pumps in order to pump water from a water source to a reservoir as shown in Fig. 5.6.

The pump powered by arrays of solar panels is either submerged underground or located on the surface depending on the water source that feeds the reservoir. This water source can be a well, borehole, lake, or a river. The pump runs only at daytime and does not use any battery. The water is pumped onto a raised water reservoir and then gravity distributes it to a low pressure drip irrigation system. Energy storage is determined by the height of the column. The size of the system depends upon the availability of water and local evaporation.

5.4.3 Standalone PV System for Livestock Watering

Delivering water to livestock in the aftermath of a disaster and in remote areas with no access to electricity is an important task. The water to be delivered is often obtained from surface sources such as ponds, rivers, streams, dugouts or underground sources such as a well. The traditional practice of delivering water to the livestock by allowing them to congregate in and around the water source has many disadvantages such as depositing animal waste near the water, destroying the vegetation, exposing the soil to erosion, and causing pollution. In order to overcome these problems, it is desirable to move water via a pipeline from the water sources to a different location and rotating the drinking areas along with the pastures. Additional benefits for moving water from surface sources via pipes is to establish vegetated riparian zone buffers. This also supports soil stability, reduces sediment loads, and improves water quality while enhancing wildlife habitat.

The use of fixed solar-powered standalone pumping systems for livestock watering is one of the most common water pumping application and the best

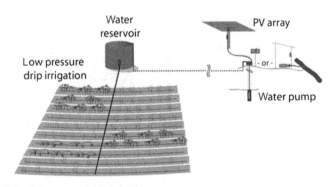

FIGURE 5.6 Solar-powered drip irrigation system.

choice of using diesel, windmill, sling, etc., considering cost, environmental concerns and reliability. Fixed solar-powered standalone pumping systems are environmental friendly, and require less maintenance than the use of windmills, gasoline, propane, or diesel generator. These systems are simple to design in which water is pumped mostly from a bored well into a holding tank near the well for consumption by the livestock. In larger systems, there may be more than one holding stock tanks located far away from the main holding tank at a high elevation. These multiple stock tanks may be gravity fed from a main holding tank.

The pump for livestock water delivery system can be operated by using either an electronic float switch or standard pressure switch. Recharged pressure tanks are commonly used with home well pumps. The recharged pressure tank prevents the continuous on/off cycling of the pump when livestock drink from a nearly-full watering tank. When the float valve closes in a recharged pressure tank system, the pump continues to run until the pressure tank is charged with water at the preset off-pressure. As the level in a near-full tank fluctuates when animals are drinking and the float valve opens and closes, water is supplied from the charged pressure tank and the pump does not cycle. When animals drink enough to lower the water level and the float valve remains open, the pressure tank water charge is exhausted and the pressure switch then turns on the pump. A check valve placed in line upstream from the pressure switch location prevents the water line from draining when the pump is not operating.

To avoid the cost of electronic shutoff controls and the apparent pump controller failures caused by them, it is recommended to use a significantly larger storage tank possibly large enough to hold the volume that the pump is capable of conveying during a 5–7 day period. The cost differential, especially when using polyethylene storage tanks are located several hundred feet away from the pump, will typically cover the increased cost of a larger storage tank and provide the benefit of a larger reservoir.

Solar-powered pumping systems for delivering water from surface sources or wells can be mobile and allows the systems to be moved to different locations. For mobile solar-powered pumping systems, the PV panel can be mounted on a trailer and set up on-site with the appropriate tilt for the panels. During the transit, it is advisable that PV panels be taken off the rack and secured in protective padding to keep them from being damaged. In the case of moving livestock several times a year to the area in the proximity of water sources, a well-designed mobile system can be a substantial saving. The reservoir in this system may be just a ditch or cattle trough set on the ground.

5.5   STANDALONE PV SYSTEMS FOR WATER PURIFICATION

Fixed standalone PV system can also be used for water purification in the aftermath of a disaster or in remote areas which have no access to electricity. According to a 2007 World Health Organization (WHO) report, 1.1 billion people lack access to an improved drinking water supply and many suffer from waterborne illnesses. The WHO estimates that 94% of diarrheal cases are preventable through modifications to the environment, including access to safe water. In remote areas, a large proportion of people without access to the clean water very often live in close proximity to the water source, which may happen to be the sea, rivers or underground waterholes. In many cases the water that surrounds them is contaminated or salty and needs to be purified. Only 1% of the planet's water is suitable for human consumption. The majority of the other 99% is salt water and the remaining water is solid frozen in ice blocks, iceberg and glaciers. However, most of the purification systems run on electricity that is nonexistent in many remote areas or areas hit by natural hazards. The goals of water purification systems design include supply of safe water, free from impurities and disease-carrying bacteria and organisms, at low cost with high reliability. There should be an option of being mobile and it could be used in areas with no access to electricity including disaster-stricken areas without clean water.

In remote areas with no access to electricity, a standalone solar-powered solution would be an ideal solution. Solar-powered water purification systems are systems that integrate solar electricity and water purification. In this method the purification happens in a water filter as it is transported (pumped or pushed) by a solar-powered pump. Photovoltaic panels generate power for a battery that pumps the water through a filtration and purification. The following solar-powered filtration methods are commonly used for water purification:

- Standalone PV system for UV sterilization
- Standalone PV system for reverse osmosis water purification
- Standalone PV system for ultrafiltration

Many other methods for purification of water are used where instead of using PV panels for powering a pump, the sun's energy is used directly to purify water for drinking and household purposes. These methods include boiling and pasteurization using solar collectors, and slow sand filtration. Nearly all of these methods are used during the day which limits their usage even when it is cloudy. In rural areas, boiling is the most common water purification method, but it requires the burning of wood or kerosene oil which can be hazardous and increases greenhouse gas emissions. It is becoming more common to use solar energy to power water treatment plants. In the case of community-based systems, the water must be stored to avoid recontamination from bacterial growth.

5.5.1 Standalone PV Systems for Ultraviolet Sterilization

In remote or disaster-ridden areas without electricity, solar-powered UV light for sterilization kills bacteria and viruses on exposure even when there is no sun or during a cloudy day. UV sterilization uses power generation consisting of PV panels, batteries, a charge controller, and safety controls. The advantage of UV sterilization using PV power generation is that excess power can be stored in the batteries for days when cloud cover limits the sun energy. Ultraviolet light for water sterilization is produced by an ultraviolet lamp that produces UV-C light at a wavelength of 254 nm with a minimum dosage of 40,000 µW/cm^2 per NSF/EPA (NSF/ANSI Standard 55–2002 Ultraviolet Microbiological Water Treatment Systems). The UV light, which is transmitted and reflected through the water, passes through the sterilizer and results into water free from harmful pathogens and rendered safe to drink. The treated water is completely unchanged to taste, odor, mineral content or other properties. Ultraviolet sterilization is very effective for treatment of water to be consumed at a single point of use, but not for multiple outlets or piping systems. Ozone systems are often recommended for distribution and bottling of water. Water that is not completely clear is not effectively sterilized by UV, as particles that may be present can protect pathogens "hiding" behind them from the UV light. Water or fluids to be treated by UV should be prefiltered to 5 µm to ensure that all particles surfaces are exposed to UV light.

5.5.1.1   Example of Standalone PV System for Ultraviolet Sterilization

Aqua Sun International has developed range of solar battery-powered water purification systems using ultraviolet sterilization to purify water taken from lakes, wells, rivers and from runoff. The application of solar power to water purification system makes it possible to provide clean water to remote villages in developing countries, disaster relief areas, army deployments, or remote areas in developed countries with no access to grid. Aqua Sun International's water purification system currently has over 2000 installations throughout the world and is distributed through various NGOs, UN agency, charitable organizations, churches, government entities, and the US State of Defense. These systems varies from portable (suitcase) containing one solar panel to stationary wall-mounted containing one or two solar panel systems. This system is approved for air shipments inside and outside continental United States. The Aqua Sun Foundation is a US charitable 501(c)(3) organization which is eligible to receive tax-deductible donations. The foundation donates drinking water purification system for disaster relief purposes and countries in need of clean water for survival.

In a three-step water purification process, the first step is to run the contaminated water through a prefiltration system to remove any sediment present in the liquid including bulk of organic contaminants. The water is then passed through another filter based on carbon block polishing. This filter will reduce fine sediments, other chemical and biological contaminants, including cysts, *Cryptosporidium, Giardia lamblia,* bad taste and odors, chlorine and chlorine byproducts, and heavy metals down to 0.5 µm in size. The third step involves extermination of bacterial, cystic and viral contaminants since purification using carbon filtering removes the majority of bacterial, cystic and viral containment. This will be achieved by using high intensity ultraviolet disinfection system that has been demonstrated to disinfect water up to 99.999999% for most cystic, bacterial and viral microorganism including cholera, dysentery, infectious jaundice, hepatitis, enteric fever, *Escherichia coli*, diphtheria, and many other microorganisms. Solar panel powers a self-priming water pump rated between 20 psi and 65 psi and an ultraviolet lamp whose output is rated at 16,000 μW seconds/cm². The main filtering process to eliminate particulates is achieved by a minimum of 20 psi provided by PV panels which are also used to eliminate bacteria and pathogens using ultraviolet purification.

In order to meet diverse needs, Aqua Sun International manufactures a range of systems including a portable "Responder S," a cart-mounted "Outpost S," and eight different kinds of wall-mounted stationary "Villager" systems. All of these systems have the capabilities of producing cleaner and safer drinking water from any lake, stream, pond, creek, well or fresh water source using a solar-powered battery at a different rate.

"Responder S," is a portable system which produces 3.70 L (1 gallon) per minute or 370 L (100 gallons) per day using solar-powered battery. Alternatively, for continuous water production, a 15 feet long, 12 V electrical power cord with cigarette lighter plug and battery jumper cable clips can be plugged into any vehicle or any 12 V electrical power supply source. It can be used for everyday usage instead of traveling miles to clean source of water.

The "Villager" water purification is a stationary, wall-mount system powered by a 12-V solar-charged battery for remote areas with no access to electricity or unreliable electricity. There are eight systems in this series which differ in size, the number of filters used, and the amount of clean water produced. These systems are mounted on aluminum powdercoated frame with mounting flange and 4 bolt hole style design.

The basic system in this category, S-1, is shown in Fig. 5.7 is powered by one PV panel of 10 W. It uses two filters, a sediment filter, and 0.5 μm carbon block filter. Disinfection is achieved by ultraviolet light. The water filtering system produces 370 L (100 gallons) of clean water per day. The PV panels measures 17 inches in length, 14 inches in width and 2 inches in width. The shipping weight without replacement parts and battery is 30 lbs.

FIGURE 5.7   Villager S-1 solar-powered water purification system for remote applications.

The largest Villager S12-4 system has three filters, washable/reusable debris filter, a sediment filter, a 0.5 µm carbon block filters and ultraviolet bulbs for disinfection. It produces 63,936 L (17,280 gallons) of clean water per day. Village S12-4 has two solar panels each measuring 38 inches in length, 27 inches in width and 2 inches in thickness. The shipping weight is 80 lbs.

The Aqua Sun's Outpost S12-4 is a lightweight three-wheel standalone solar-powered system which does not require any outside electrical power. A 100 W solar panel as shown in Fig. 5.8 is attached to the top of the unit which charges a 12 V deep cycle battery underneath it.

The solar-charged battery powers a water pump and UV light to produce 3700 L (1000 gallons) of purified water per day based on fully charged solar/battery. The water purification system uses a washable/reusable filter followed by a sediment prefilter and three-stage 0.5 µm carbon block polishing. The unit is made of aluminum frame with eight handle grips for easy lifting into the back of a vehicle. The empty weight of Outpost S12-4 is 280 lbs, and has the capabilities of using an auxiliary power source of 12 V vehicle or generator. Outpost S12-4 comes with many optional components such as a surface or submersible solar water pump, AC to DC power supplies, outpost chlorinator, water tanks and 4-faucet pressurized filling station for automatic operation and steady water flow.

5.5.2 Standalone PV System for Reverse Osmosis Water Purification

Reverse osmosis is a membrane separation process for removing solvents from a solution using a crossflow process. This process is the opposite of the natural osmosis process of water which depends upon a semipermeable membrane through which pressurized water is forced. For purification of water, the feed

FIGURE 5.8   Ultraviolet light water purification system for remote applications.

water is pumped on one side of the semipermeable membrane at a pressure high enough to exceed osmotic pressure to cause reverse flow of water. If the membrane is permeable to water, but impermeable to dissolved contaminants, the pure water will cross the membrane. This crossflow of pure water increases the concentration of dissolved contaminant in the remaining feed water resulting in increase in osmotic pressure. As molecules of contamination are physically larger than water molecules, the membrane blocks the passage of contaminant particles. The end result is purified water on one side of the membrane and highly concentrated water with impurities on the other side. In addition to impurities, this process will remove a select number of drinking water contaminants, depending upon the physical size of the contaminants. The pores in reverse osmosis membranes are approximately $0.0005\,\mu m$ in size while bacteria and viruses size are 0.2 and $1\,\mu m$ respectively. There are two types of commonly used reverse osmosis membranes for home water purification products: Thin film composite TFC and cellulose triacetate (CTA). A TFC membrane is a multilayer membrane in which an ultrathin semipermeable membrane layer is deposited on a finely-microporous support structure. It has considerably higher rejection rates for the contaminants than CTA membrane but is more susceptible to deterioration by chlorine. As a result, a reverse osmosis system often uses a prefilter of activated carbon.

A solar-powered reverse osmosis water purification system uses a solar-powered pump to push the fluid through the energy-saving TFC membrane which is highly resistant to waterborne microorganisms, and shown to have extraordinary purification efficiency for all kinds of contaminants. A high-efficiency design can be coupled directly to PV panels to obtain the maximum energy transfer between the solar panels and the reverse osmosis pump. The output from a solar-powered unit will generally be limited to the hours of sunshine. However, an optional rechargeable battery-pack can store excess electricity produced during peak hours of sunshine, allowing operation for a few hours even at night for that extra, emergency drinking water supply.

5.5.2.1 Example of a Standalone PV System for Reverse Osmosis Water Purification

Blue Spring Inc. has developed the EC-S series of solar-powered reverse osmosis water purification systems that can produce WHO-quality drinking water from contaminated water sources such as rivers, lakes, ponds, wells as well as low-salinity brackish water sources. These reverse osmosis based systems are capable of removing a wide range of impurities including arsenic, bacteria, dissolved salts, humic acid mercury, nitrates, organics, viruses, and others with greater efficiency. The Blue Spring EC-S series reverse osmosis water purification systems can be temporarily used for victims of weather-related disasters on an emergency basis, or for a household consumption in remote areas.

The solar-powered purification system as shown in Fig. 5.9 consist of PV panels, reverse osmosis membrane, a prefilter, a pump and an optional rechargeable battery pack.

Solar panels convert the sunlight into electrical power at places where there is no electricity or the supply is unreliable. The amount of electricity produced by the solar panels depends on the size of panels and amount of incident sunlight. This electricity is used to push the water by solar-powered high pressure pump through TFC reverse osmosis membrane. In case of cloudy days and nighttime a backup battery is used to power the process of filtering. A fine grade prefilter is also used to ensure removal of sand and other particles which extends the operating life of the reverse osmosis membrane. Media filters are also available on optional basis for purifying cloudy, hazy, and muddy sources of water.

Water is stored in a tank from where it can be distributed to the household or community. These system use highly efficient PV panels, and uses state of the art technology in reverse osmosis membranes that minimize power consumption and are economical to run. The combination of the use of highly-efficient PV panels and low-pressure TFC membrane makes it so efficient that it works in sunlight as low as 30% of the peak value during the day. This results into production of 700 L of purified water per day per square meter (17 GPD/foot^2) of the earth's surface. The PV panels used in these system have a long life of 25–30 years and can be mounted on the roof or ground for capturing maximum sunlight. For a longer life, parts of the system are made from welded aluminum frame and the materials that come in contact with the water are corrosion-resistant brass or specialty polymers. The production of water also depends on the amount of sunlight available in the area of application and salinity of the incoming water. The EC-S series of systems can be accurately used for water salinity up to 3000 mg/L.

FIGURE 5.9   Solar-powered reverse osmosis water purification system.

For higher water salinity, Blue Spring has manufactured five solar-powered systems, the SW-S series, which can be used for brackish water and sea water desalination systems. These systems vary in size and produce clean water output of 300–6800 L/day (80–1800 GPD) depending on the model used. The first two models, SW-80S and SW-1MS, can be carried by two people and set up in a few minutes. The fresh water production capacity of these systems are 300 L/day (80 GPD) and 570 L (150 GPD) respectively and can easily supply clean water for one or two homes, a restaurant or tourist location near the coast or an island. These portable drinking water systems can be appropriately used in the aftermath of weather-related disasters in the coastal areas as well as for fishing and sailing boats. The remaining three models, SW-2MS, SW-3MS, SW-7MS, are standalone systems with fresh water production capacity of 1300 L/day (340 GPD), 2500 L/day (660 GPD) and 6800 L/day (1800 GPD) respectively and are suitable for coastal communities, islands, off shore oil rigs, as well as for ocean liners, and cargo and naval ships.

These systems vary from the portable model EC-450S purifying 120 GPD to EC-30MS standalone purification system purifying 8000 GPD. EC-450 S weighs 112 lb (with PV panels) and can be easily carried and set up by one person. Its application include emergency water purification for victims of weather-related disasters and source of clean drinking water for single family in the remote areas having no access to electricity. The largest system EC-30MS is a special order which weighs nearly 1984 lbs (with PV panels) and is used as a standalone system for supplying fresh water to the victims of weather-related disasters and remote communities of up to 160 people.

Other systems in series include;

- EC-1MS: weight 147 lbs, output capacity 320 GPD
- EC-2MS: weight 188 lbs, output capacity 560 GPD
- EC-3MS: weight 223 lbs., output capacity 720 GPD
- EC-4MS weight 314 lbs, output capacity 960 GDP
- EC-6MS weight 485 lbs, output capacity 1500 GPD
- EC-12MS: weight 765 lbs, output capacity 3000 GPD
- EC-18MS, weight 445 lbs, output capacity 5000 GPD

5.5.3 Standalone PV Systems for Ultrafiltration

Ultrafiltration is a type of membrane filtration where hydrostatic pressure is applied to separate materials from water using a semipermeable membrane pore size of 0.002–0.1 μm. This is a low pressure membrane process which needs approximately 30–100 psi to exclude particles based on size and does not need any chemicals. The use of a hollow fiber membrane provides reliable rejection of microorganisms and viruses with low operating costs. This is achieved by feeding water inside the shell, or in the lumen which allows suspended solids and solutes of high molecular weight to be retained and water and low molecular weight solutes pass through the membrane. This process excludes particles as small as 0.01 µm, including bacteria, viruses, and colloids which provides a reliable, stable, and consistent water that meets the water quality Standards around the world.

5.5.3.1   Example of Solar-Powered Ultrafiltration System

GenPro Energy Solutions has manufactured three solar-powered ultrafiltration systems for water filtration and purification that can be used in remote and disaster-ridden areas. These systems differ in water producing capacity, mobility, and size as explained below:

GENPRO 1000, as shown in Fig. 5.10, is the largest of three models: a standalone trailer with a maximum water-producing capacity up to 37,854 L (100,000 gallons) of microbiologic safe drinking water per 24 hour timeframe.

The GENPRO 1000 is powered by four retraceable/adjustable solar panels of 190 W each (maximum 760 W) which charges the battery bank with a normal solar daytime frame based on the location of the solar panels. The system uses a fiber membrane of 4 inches in diameter and 40 inches in length with an estimated flow rate of 1300–3900 GPD. It also uses a 20″ spun depth sediment filter in addition to a high-flow 1 µm carbon block postfiltration to reduce taste and odor. The normal feed pressure of the system is 35+ psi and maximum operating pressure is 100 psi generated by DC 3-phase water pump with 75 feet draw. The charged batteries can supply power to run the system up to 24 hours with full water production during night-time or in the absence of electrical power. It can also be powered by other sources such as a local grid, diesel generator, or wind turbine. It weighs 1179 kg (2599 lbs) and can be easily transported by a trailer or by air. The system is easy to set up and has only minimal cost to operate and maintain.

GPXM 2500 is a portable system that produces 10,000 L/day (2645 GPD) of drinking water using a combination of filters including hollow-fiber ultrafiltration membrane, high-capacity gradient depth prefilter and granular activated carbon postfilters. GPX 2500 is powered by a solar-powered DC power source which drives an external submersible 24 V DC pump to draw water from lakes,

FIGURE 5.10   Solar-powered ultrafiltration system.

wells, and streams, bore holes or holding tanks. The system is light (75 lbs) enough to be deployed by one person and is designed for field use emergency conditions. It is easy to transport as all system components are built into the rolling case and may be checked as baggage on most commercial airlines.

The GP XSS 1440 is a portable system with a manual backup hand pump that uses hollow fiber ultrafiltration membrane to remove 99.999% of bacteria, cyst, and viruses. It can produce up to 5450 L/day (1440 GPD) with 3 psi of water pressure using a back-flushable sediment filter as a prefilter and 0.5 μm carbon as a postfilter. The GP XSS 1440 is powered by an 80 W foldable solar panel and uses a 24 V DC electric pump. It weighs 124 lbs and can be transported by two persons to purify water from lakes, ponds, and streams. The system is easy to maintain and is specifically designed for water purification in the aftermath of a disaster. It is currently deployed in Haiti and southern Sudan.

5.6 DIRECT SOLAR WATER TREATMENT FOR PURIFICATION OF WATER

Direct solar water treatment for purification of water is suitable for areas where there is no electricity and no water infrastructure. It is a low-technology solution in which energy and heat is captured from the sun and used for cleaner water for human consumption. This technology is inexpensive, easy to understand, and easy to implement. Direct solar treatment provides an alternative way of providing clean and safe drinking water to remote areas and areas after disasters. There are three main types of direct solar water treatments:

- Solar water disinfection
- Solar pasteurization
- Solar water distillation using solar stills

5.6.1 Solar Water Disinfection

Solar water disinfection (SODIS) is a simple method to purify drinking water by making use of the sun's UV rays which will inactivate pathogens that can cause diarrhea. It was first introduced in 1980 by Aftim Acra et al. for disinfecting drinking water and oral rehydration solutions based on a batch system using clear or blue-tinted containers made of glass or plastic. SODIS was further developed in 1991 for developing countries by the Swiss Federal Institute of Science and Technology and is recommended by the WHO as a viable method for household water treatment and safe storage.

SODIS kills many water borne pathogens related to many health issues including cholera, diarrhea, hepatitis, fever, polio, typhus and others. These pathogens include bacteria, viruses, and the giardia and cryptosporidia parasites. The process of disinfection is achieved by the radiation of UV-A rays present in the sunlight which possibly damages the respiratory system of pathogens. This process works well even when the temperature is cool. It is however important that the water to purified is relatively clear for the UV-A rays to penetrate and disable the functions of the pathogens. It is also important for ensuring proper purification of water that waters of higher turbidity must be filtered before SODIS treatment. Contaminated water is filled into clear PET (polyethylene terephthalate) bottles and exposed to full sunlight. During the exposure, the UVA (ultra violet-A) radiation (wavelength 320–400 nm) of the sunlight destroys the pathogens. Bottles of 2 L (70 ounces) or less should be filled three quarters to permit enough oxygen for the process, then shaken for 20 seconds, and finally filled completely. When the water is turbid (turbidity higher than 30 nephelo-metric turbidity units), it must be filtered prior to exposure to the sunlight.

This process of purifying water involves exposure to the UV rays and temperature; the higher the temperature the water reaches in the bottle, the faster the disinfection process. If it is sunny or up to 50% cloudy, the water is disinfected in 6 hours of exposure. If it is 100% cloudy then the bottles need to be exposed for 2 days. If water temperature goes above 50°C the disinfection process is complete in 1 hour. After this time period, the water is disinfected and ready to drink. These simple calculations of sunlight and temperature make this method of water purification very accessible.

5.6.2   Solar Pasteurization

Solar pasteurization is the process of disinfecting water using the sun's heat in the same way that boiling water kills harmful bacteria. This was discovered by Louis Pasteur in 1880 and was used to kill microorganisms or microbes by heating water to a required temperature. Dale Andreatta, who developed the current Water Pasteurization Indicator (WAPI), discovered in 1994 that the temperature of water has to be raised to 65°C for 6 minutes to kill any harmful microorganisms. A WAPI is a simple thermometer-like device that indicates when water has reached pasteurization temperature and is safe to drink. Solar water pasteurization has been shown to kill bacteria, viruses, protozoa, and worms. Water can be heated by solar cooker, solar puddle, or solar heater. The details of these methods are given below.

- *Solar box cooker*

Solar box cookers use solar energy to sterilize water which is a simple method of achieving solar pasteurization to eliminate microorganisms from water at the temperature of 150°F (65.5 C°). A solar box, shown in Fig. 5.11, consists of an insulated box constructed from wood or cardboard with a glass or plastic lid. The inside surfaces of the box must be black.

A covered pot (preferably black) of water is placed inside the box and should remain in the box until pasteurization temperature is achieved for few minutes. The solar box collects heat from the sun and heats the water to the required 150°F and can purify 1 gallon of water in about 3 hours on a sunny day. Large

FIGURE 5.11 Solar box cooker. *solar cooking.org.*

clear plastic containers can also be used with the sides taped or painted black. Similar to the solar box, the water is heated to the proper temperature, and the water becomes drinkable. As there is different types of solar cooker designs, one should choose the one suitable to their requirements.

- *Solar puddle or pond*

A solar puddle or pond can also be used as an alternative to a solar cooker when the materials for solar cookers are not available. This is obtained by digging a hole in the ground of approximately 4 inches deep and 3 feet wide and insulating it by materials such as leaves, grass, paper, straw, and other solid materials. The insulation is then covered with layers of clear and black plastic extends out of the sides of the hole as shown in Fig. 5.12. The water to be pasteurized is placed on top of this layer, and the entire hole is covered by two other layers of plastic separated by a spacer cover. The sun heats the water to the critical temperature, after which the water can be siphoned out of the hole transferring the water from one plastic container to another. The solar puddle will have the same radiation effect as a solar cooker to bring the water to pasteurization temperature although water must be added to the hole each day.

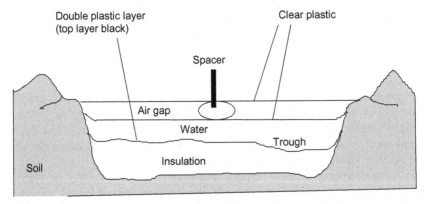

FIGURE 5.12 A basic schematic of a solar pond. *Dale Andreatta, http://solarcooking.org/pasteurization/solarwat.htm.*

- *Solar Heater*

Water pasteurization can also be obtained by solar heating using flat plate collectors enclosed in a serpentine pipe configuration. The advantage of using this configuration is to produce higher temperature change per unit area. The solar heater is estimated to produce about 11.3 L/hours at an insolation of 600 W/m² for the calculated area of an individual serpentine collector of 2 m². This will require about 31 collectors based on an average operating day of 6 hours at this insolation level. This method of water pasteurization has many advantages including minimal use of electrical components, limited filtration needs, and easily available spare parts. This solar heater has a higher cost and limited operation when there is no sun. Multiple locations and higher temperatures may also foster bacteria growth in the storage tank.

5.6.3   Solar Water Distillation Using Solar Stills

Solar water distillation is a low-technology solution for water purification. It uses a solar still to condense pure water vapor and remove metals, bacteria, salt, and other impurities. Although this method of water distillation was used thousands of years ago to produce salt by distilling seawater, the construction of a large scale still was first reported in 1872 in Las Salinas, Chile to supply drinking water to a mining company.

Solar stills as shown in Fig. 5.13 are constructed by using a glass panel on top of a dark-colored (usually black) basin that holds the water.

A plastic sheet is used to form a trap to collect water, and the glass sits at an angle above it. The sun radiation permeates the glass and the dark-colored basin heats the water that evaporates and condenses on the glass. The condensed droplets trickle down the angled glass and collects in a trough, which runs to a collection basin. During this process, the recondensed water is free of impurities and contaminants are left behind in the basin. This process of water purification can be used instead of boiling water and is useful in places where rainwater is used as the main source of drinking water. When water contains pollutants, it should be filtered before being distilled. Solar stills last for many years and this method of water purification requires only a periodic cleaning to prevent buildup of dirt or mold.

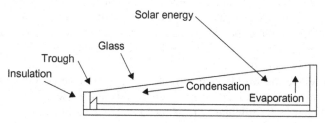

FIGURE 5.13   Solar water distillation using solar stills. http://www.solaqua.com/solstilbas.html.

The intensity of solar energy falling on the still is the most important parameter affecting the production of water. The daily distilled water output (Me in kg/m^2 day) is the amount of energy utilized in vaporizing water in the still (Qe in J/m^2 day) divided by the latent heat of vaporization of water (L in J/kg). Solar still efficiency (n) is the amount of energy utilized in vaporizing water in the still divided by the amount of incident solar energy on the still (Qt in J/m^2 day). These can be expressed as:

$$\text{Solar still production : Me} = \text{Qe / L}$$

$$\text{Solar still efficiency : n} = \text{Qe / Qt}$$

The general operation is simple and requires facing the still towards solar noon, putting water in the still every morning to fill and flush the basin, and recovering distilled water from the collection reservoir in case of glass bottles.

5.7 CASE STUDY OF FIXED STANDALONE PV SYSTEMS FOR MOBILE COMMUNICATIONS IN REMOTE AREAS

Mobile communication has become a widely-used mode of communications both in the developing and developed world. Nearly three billion people in the world use cellular communications and this is replacing the use of landlines in many places in the developed world. Mobile communications is a form of cellular communication that requires a hand set and a base station. A base station is a standalone wireless communication system and is used to communicate as part of wireless telephone system such as GSM or CDMA cell sites. Base stations need to provide all day and night operations and are installed in urban areas as well as in remote areas including deserts, mountains, and islands. In remote areas without electricity, a standalone power system is required which must be reliable, cost effective, and easy to maintain. Base stations are not manned and as a result require high level of reliability and longer lifespan of power supply which is used to power equipment. According to one estimate, there are about 5 million cell phone towers in the world and 64,000 of them are off-grid and run by diesel power. In the developing countries, the numbers of cell towers are increasing at the rate of 50,000 per year as cellular networks are extended to the remote areas. The vast majority of these cell towers are run by diesel power because they are far from the grid or suffer from unreliable grid power.

The use of diesel generators to power base stations can be expensive, less reliable, cause air pollution, and require regular maintenance. The new approach is to power base stations with renewable energy which can offer clean, reliable, and cheap electricity. This can also help to combat climate change. In order to bridge the worlds of renewable energy and mobile communications (base stations) and optimize overall system efficiency, the GSM Association (GSMA) launched a program called Green Power for Mobile in Sep. 2008. The purpose of the program was to promote the use of renewable energy by the mobile phone industry. The goal of this program was to achieve 118,000 new, existing, and off-grid base stations to be powered by alternative energy by 2012. In addition, the program would save up to 2.5 billion liters of diesel a year and reduce annual greenhouse gas emission by up to 6.8 million tons. Pike Research predicted in 2010 that renewable energy will be used to power 4.5% of the world's mobile base stations by 2014, which is an increase from 0.11% in 2010. In developing countries this prediction is 8% which is much higher than developed countries. In 2013, Navigant Research predicted that revenue from off-grid power for mobile base stations will top $10.5 billion by 2020. This upward trend in the market for green base stations for mobile communication is the result of rising energy costs, government policy initiatives and concern for environment. These green base stations which use a combination of solar energy, wind energy, batteries, and fuel cells could become much more prevalent within 10 years.

5.7.1   Solar-Powered Base Stations

A solar-powered base station as shown in Fig. 5.14 consists of a PV powering unit, a base station and a cooling unit. The base station uses radio signals to connect devices to network as a part of traditional cellular telephone network and solar powering unit is used to power it. The PV powering unit uses solar panels to generate electricity for base stations in areas with no access to grid or areas connected to unreliable grids. The PV powering unit shown in Fig. 5.14 consists of photovoltaic arrays, battery packs, an inverter, a charge controller, junction boxes, and cables to connect various components of the system.

The photovoltaic array converts sunlight into 48 V DC to power communication equipment of the base station. The DC output of photovoltaic module is also converted to 110/220 V AC using an inverter for powering air conditioning of the base station. Photovoltaic module normally uses silicon monocrystalline or polycrystalline cells with a typical output voltage of 0.5 V. In order to generate 48 V DC, two solar panels consisting of 72 PV cells should be connected in

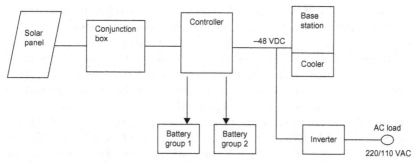

FIGURE 5.14   Solar-powered base station.

series to obtain output voltages ranging from 43.2 to 56.4 V. Power for air conditioning is obtained by converting DC output of solar panels into 110 V/220 AC using an inverter. A photovoltaic module should be placed at a certain inclination to maximize the solar contribution to a particular load based on load consumption, climate, temporal profile, and latitude. In general, a PV module with tilt angle equal to the latitude of a location receives maximum insolation. However, in some locations the morning and afternoon insolation is not symmetric as the winter may be cloudier than summer. In order to maximize the incident insolation and reduce the number of batteries, it is usual to use inclination of 10–15° more than latitude during the Winter and inclination of 10–15° less in the Summer. Zinc coated steel bracket are used to support the PV modules which fix the modules at a certain tilt angle. A bank of batteries are used to power the base station when the solar power is not enough at night or during the rainy and cloudy days. The batteries are also charged by solar panels when not used to provide power and its capacity depends on load, backup time of the battery, discharge depth, and safety requirements. A charge controller is connected between the solar panels and the batteries to ensure that the maximum output of solar panels is used to charge the batteries without overcharging or damaging them. Depending on the size of base station and its traffic, the base station may also have another sources of power such as a diesel generator, wind turbine or biofuels.

The base station is a transceiver and acts as an interface between a mobile station and network using microwave radio communication. It consist of three part elements: one or more transceivers, several antenna mounted on a tower or building, power system, and air conditioning equipment. A base station can have between 1 and 16 transceivers, depending on geography and the demand for service of an area. The base station antennae are mounted on tall towers because it is easier to stay in communications with mobile phone users and avoid obstacles such as tall buildings, trees, and hills that stand between the base stations and mobile phone. Usually, there are three antennas installed to cover the specified region, where two of the three antennas are used for receiving and transmitting. The two antennas for receiving are used to choose stronger signals which are much weaker than the transmitting signals. The power of the signals depends on the size of the cell which the base station uses to cover a specific area. The size of cell ranges from 1 to 20 km (macro), 400 m to 2 km (micro), 4 to 200 m (pico) and around 10 m for femto cells. Macro cells covers the largest area among all the cells and are generally used for rural or remote areas. The power needed to run the base station depends on the size, the coverage area, and the technology used by the network. It also depends on the number of calls at that time which is lower during the night time than at daytime. For instance, a typical 3G base station consumes about 500 W of input power to produce about 40 W of RF power making it the average annual energy consumption of 3G base station around 4.5 MWh. A 3G network with 12,000 base stations will consume more than 50 GWh in a year. Power consumption for base stations will be much greater in countries with large geographic area and populations. In China, the mobile phone market has over 580 million subscribers and is serviced by 5,000,000 GSM, 200,000 3G and CDMA base stations. Such power consumption not only produces large amount of CO_2 emission but also increases the system operating energy and makes it more expensive.

5.7.2   Benefits of Solar-Powered Base Stations

The use of solar energy to power base stations will not only reduce the cost of operation, but also allow deeper penetration of mobile networks in the remote areas. The cost advantage of solar energy is more pronounced with the continuous expansion of photovoltaic cells and exhaustion of coal and oil, and its impact on environment. With the extension and upgrading of the world's telecommunication network, rural communication services are becoming more important especially with the popularity of high bandwidth mobile services to connect people living or travelling in remote areas. It is predicted by InStat that there will be over 230,000 solar-powered or wind-powered cellular base stations in developing countries by 2014. In remote areas of developing countries, it will also be economical to use solar or wind to power base stations rather than constructing high voltage power grids. Regions like Africa, Australia, Central America, Middle East, South Asia, and South East Asia have abundance of sun that can be used to generate electricity to power cellular or mobile communication equipment. Some of the benefits are detailed below.

5.7.2.1   Cost Effectiveness

Solar-powered base stations do not need power lines to the areas that do not have existing power, hence no cost will be incurred in buying and running a generator to the base station. Once the solar-powered systems are installed, they require little or no maintenance which further reduces running expenses over the lifespan of the base station.

5.7.2.2   Reliability

The solar-powered base stations are standalone systems and are independent of any utility's grid power outages. Such a system would provide a more reliable communication network especially in developing countries where the grid is not reliable and suffers many power outages.

5.7.2.3   Flexibility of Location

The installation of solar-powered base station does not specifically require a fixed position making the system flexible. It can be relocated as needed. The only constraint on these systems is whether or not location is available or not, unlike grid powered base stations which require adequate power source at a particular location for maximum coverage. Flexibility in location is extremely helpful in remote areas where power availability is low.

5.7.2.4 Environmentally Friendly

Solar power uses the sun as a renewable energy and does not cause any pollution. The utilization of sun's energy in powering the base station would lead to reduction in the ecological footprint and are more environmentally friendly.

5.7.3 Examples of Solar-Powered Base Stations for Disasters and Remote Areas

Solar-powered cellular base stations have been used both in developed and developing countries when the power from the grid is disrupted due to weather related disasters. Solar-powered base stations are also used in remote areas far removed from the grid or in areas where the power supply is unreliable and expensive to install. Below are some examples of the use of solar-powered base stations for disaster-struck and remote areas.

In Vermont, United States, a Canadian border town of Norton maintained communications with the outside world by using a solar panel and battery system on a cell tower during flooding from Tropical Storm Irene in 2011. The US department of Economic Development awarded a $1.6 million grant to help deploy resilient cellular broadband services to nine towns that were damaged by Tropical Storm Irene and other flooding under the Vermont Resiliency Project. The purpose of the project is to partially use solar-powered cellular and wireless equipment to ensure the continuity of communication services during weather-related emergencies. This is achieved by expanding cellular service on roadways and creating Wi-Fi hotspots in nine towns affected by weather-related disasters. The Wi-Fi hotspots will be able to operate during power disruptions with the help of solar and battery system and will provide both cellular and internet services. The project plans to expand cell service on 120 miles of area covering nearly nine counties in Vermont.

In the aftermath of the 2011 Great East Japan Earthquake, NTT DOCOMO Inc. of Japan has developed and field tested three mobile network base stations powered by solar panels, high-capacity rechargeable batteries, and green power controllers. This arrangement will be used to establish 10 green base stations in and around Tokyo, Kanagawa, and Yamanashi. The green base station uses solar panels to generate electricity and store it during daytime by charging high-capacity rechargeable lithium–ion batteries. The stored energy from rechargeable batteries will be used to power the base station during the weather-related disaster when electricity supply from the grid is disrupted. It will also store any excess power for nighttime consumption. Remotely-located green power controllers are used to manage various base station operations including enabling solar power to be stored as direct current to improve the conversion efficiency. Maximum consumption of base station is 2.0 kW and the power generated from the solar panels is 4.19 kW. The high-capacity rechargeable batteries can store between 14 and 16 hours' worth of power when energy from sun is not available. Cost of electricity can be reduced by charging the batteries with less-costly electricity during night and using it during daytime when the costs are higher. By Mar. 2015, NTT DOCOMO has successfully tested its green base station with dual source technology which is claimed to save 95% of solar and off-peak usage and provide reduced electricity consumption by more than 90%.

In the United States, less than 1% of base stations are powered by renewable energy but that figure is slowly increasing. In 2011, Verizon has 20 solar-powered base stations. T-Mobile added renewable energy to more sites and also uses it as temporary backup power which will expand in the coming years.

Alcatel-Lucent has recently installed more than 300 solar-powered wireless base stations in Sub-Saharan Africa. The success of these early projects has attracted the attention of network operators elsewhere, not just in areas without a reliable electricity grid. Several leading operators in Europe and other highly developed areas are seeking to reap environmental benefits while lowering their energy costs, and hence are currently evaluating proposals to introduce renewable energy-powered base stations.

In Jul. 2012, when a power failure disrupted the electricity available to 670 million people in India, the base stations that service nearly one billion mobile phones largely stayed operational. The majority of these base stations that had a backup diesel power are being investigated for powering via clean renewable energy including solar energy.

5.7.4 Solar-Powered Base Station in India

India is the world's largest democracy with a population of over 1.27 billion and has a growing economy. Its telecommunication network is the second largest after China, based upon the number of fixed and mobile users. The number of mobile subscribers are expected to peak to 1 billion in 2015 with the number of cellular base stations reaching up to 500,000. The problem of providing energy to remotely-located systems is a serious concern for the telecommunication industry. This problem exists particularly among the mobile telephony towers in rural areas, that lack quality grid power supply. A cellular base station can use anywhere from 1 to 5 kW power per hour depending upon the number of transceivers attached to the base station, the age of cell towers, and energy needed for air conditioning. Cellular base stations use power without any interruption and also needs maintenance. The increase in demand of power base stations from Indian telecommunication industry is a big challenge, especially in rural India.

The majority of these base stations in India use diesel as they are either far from the grid or electricity from the grid is not reliable. It is estimated that almost 70% of telecommunication towers in India are located in areas with more than 8 hours of grid outages and almost 20% are located in off-grid areas. Most transceivers in the cellular base stations are run by 48 VDC to charge the batteries and power the communication equipment. The air conditioning of the base station runs at 220 VAC. These base stations can be powered by two types of diesel generators. The first is the conventional type where 220 VAC is converted to 48 VDC to charge the batteries and power the communication equipment. The second type has a DC output which is used to charge the batteries directly and power the communication equipment. For running the air conditioning, the voltage needs to be converted to 220 VAC.

According to Greenpeace India, the Indian telecommunication industry consumed an estimated 3.2 billion liters of diesel in 2011. This will increase to 6 billion liters by 2020. Diesel generators produce substantial amount of greenhouse gases and harmful air pollutants including black carbon (BC), carbon monoxide (CO), nitrogen oxides (NOx), sulfur dioxide, hydrocarbons (HC), and particulate matter (PM). The impact of diesel combustions can have health risks not only for humans but also to air, water, and soil both locally and globally.

The Indian government, which heavily subsidizes diesel, has recently mandated that 50% of rural sites be powered by renewable energy by 2015. This decision is made because the Indian government wants to reduce greenhouse gas emissions and lessen the country's reliance on imported oil. The mandate also requires that 75% of rural and 33% of urban stations need to run on alternate energy by 2020. Such a move by India in 2015 will likely bring down the market prices in different regions of the world which are planning to use renewable energy-powered base stations. Although such a move will increase the number of cellular base stations, it will however reduce carbon emission and save a substantial amount of fuel and money.

In recent years, the telecommunication sector has shown an increased interest in the adoption of solar-powered cellular base stations due to financial benefits, accessibility to remote areas, and reduction in green gases in the environment. One obstacle of entry of solar energy to cellular base stations is an intensive power requirement of the current base stations. As a result, the electronic industry is exploring new methods to reduce the power requirements for the electronic equipment used in the base stations. The first approach is to make the base stations more tolerant to heat which will then require less power for air conditioning. The other approach is to use integrated chip technology and smaller radios to improve energy efficiency. This includes electronic units that can turn themselves off automatically when not in use. The trend of employing distributed small cells using smaller antennaes and radio that uses much less power is increasing and could be more attractive to both developing and developed countries. The industry is also using real-time adjustment to optimize power usage and monitoring renewables by integrating energy monitoring capability into base stations.

A typical Indian cellular base station running on diesel can cost up to US$14,510 per year while a solar-powered base station with battery backup costs only US$8215 per year. It is expected that while the cost of diesel fuel will exceed US$20,000 in 2020, the prices of solar technology will continue to fall to less than US$5500 per year. A recent study showed that global power consumption for cellular base stations will decline due to more efficient equipment and networks by nearly 3% annually while the cost of electricity powering these base stations will rise by 9% annually. Research suggests that electricity costs will rise faster than savings gained through more efficient equipment and networks.

In view of these benefits, the Indian Ministry of New and Renewable energy is expected to incorporate a project into India's National Solar Mission which plans setting up 20 GW of solar power capacity by the year 2022. This move by the Indian government would not only help them achieve their goal of powering 50% of their rural base stations by renewable energy but allow them to get subsidies for telecommunication sector for installing solar-powered base stations. India's decision to convert diesel to solar for their base stations will have the greatest impact on other developing countries. More than half of the base stations in these developing countries are not connected to the electric grid.

In order to design and implement a solar-powered base station, PVSYST simulation software has been used in various countries including India, Nigeria, Morocco, and Sweden. This software allows for estimation of the number of PV panels, batteries, inverters, and cost of production of energy considering the geographical and other design parameters. It is essential to use PVSYST software and get an understanding of efficient and cost effective solar-powered base station before it is built.

BIBLIOGRAPHY

[1] D. Andreatta, A Summary of Water Pasteurization Techniques. <http://www.sswm.info/sites/default/files/reference_attachments/ANDREATTA%202007%20A%20Summary%20of%20Water%20Pasteurization%20Techniques.pdf>, 2007.

[2] D. Andreatta, The solar puddle-a low cost water pasteurizer, in Annual Conference of the American Solar Energy Society, 2001

[3] A.V. Anayochukwu, E.A. Nnene, Measuring the Environmental Impact of Power Generation at GSM Base Station Sites. Published online 30 April 2013. doi:10.7770/ejee-V1N1-art479. Renewable Energies Research Nucleus, UC Temuco, 2013.

[4] Aqua Sun, Solar-Powered Water Purification System. <http://www.gpra.50megs.com/Aquasun.htm> <www.aqua-sun-intl.com>.

[5] Blue Spring EC-S Series, Solar-Powered Reverse Osmosis Water Purifiers. <http://www.bluspr.com/solar_reverse_osmosis.html>.

[6] J. Burneya, L. Wolteringb, M. Burkec, R. Naylora, D. Pasternakb, Solar-powered drip irrigation enhances food security in the Sudano–Sahel, PNAS 107 (5) (2010). <http://www.pnas.org/cgi/doi/10.1073/pnas.0909678107>.

[7] M.J. Buschermohle, R.T. Burns, Solar-Powered Livestock Watering Systems. The University of Tennessee Agricultural Extension Service. PB 1640. <https://utextension.tennessee.edu/publications/documents/pb1640.pdf>.

[8] M. Chadha, Solar-Powered Cellphone Towers in India to Reduce 5 Million Tons CO_2 Emissions, Save $1.4 Billion Every Year. Solaripedia <http://www.solaripedia.com/13/249/cell_towers_combine_solar_&_wind_power.htm>.

[9] A. Chatzipapas, S. Alouf, V. Mancuso, On the minimization of power consumption in base stations using on/off power amplifiers, in 2011 IEEE Online Conference on Green Communications, 2011. doi: 10.1109/GREENCOM.2011.6082501.

[10] Combating Waterborne Diseases at the Household Level, World Health Organization. <http://www.who.int/household_water/advocacy/combating_disease/en/>, 2007.

[11] J. Currit, S.E. Jones, Solar Cookers for Developing Countries, Brigham Young University, Provo, Utah. <http://solarcooking.org/solar-ovens-for-developing-countries.htm>.

[12] W.S. Duff, D.A. Hodgson, A Simple High Efficiency Solar Water Purification System. <http://www.invenzone.com/research_papers/a-simple-high-efficiency-solar-water-purification-system-59212851>, 2004.

[13] Eco-Sustainable Wireless Networks: Ready for Prime Time Alcatel–Lucent <http://www3.alcatel-lucent.com/wps/DocumentStreamerServlet?LMSG_CABINET=Docs_and_Resource_Ctr&LMSG_CONTENT_FILE=Other/Eco-Sustainable_Wireless_Networks_Article.pdf>.

[14] K. Fehrenbacher, Clean Energy to Power 4.5% of Cell Phone Base Stations. <https://gigaom.com/2010/07/06/clean-energy-to-power-4-5-of-cell-phone-base-stations/>, 2010.

[15] Green Solutions for Telecom Towers: Part II Solar Photovoltaic Applications, Intelligent Energy. <http://www.intelligentenergy.com/media/uploads/green_solutions_for_telecom_towers_part_2_solar_photovoltaic_applications.pdf>, 2013.

[16] GSM Association, 3 Billion GSM Connections on the Mobile Planet. <http://www.gsmworld.com/newsroom/pressreleases/2008/1108.htm>.

[17] GREENPRO Energy Solutions. Water Filtration and Purification. <http://www.genproenergy.com/>.

[18] G. Hanson, Unsafe Drinking Water, the Solar-Powered Solution. Sustainable Development International. <http://infohouse.p2ric.org/ref/40/39721.pdf>.

[19] K. Harrington, Solar Power Making Significant Inroads in Rural India. <http://www.aiche.org/chenected/2015/01/solar-power-making-significant-inroads-rural-india>, 2015.

[20] K. Heimerl, E. Brewer, The Village Base Station. <http://www.cs.berkeley.edu/~kheimerl/pubs/vbts_nsdr10.pdf>.

[21] T. Hoium, Why Industries Like Oil and Gas Are Turning to Solar Energy. <http://www.fool.com/investing/general/2015/02/28/how-solar-energy-can-helping-oil-and-telecom-compa.aspx>, 2015.

[22] D.U. Ike, A.U. Adoghe, A. Abdulkareem, Analysis of telecom base stations powered by solar energy, Int. J. Sci. Technol. 3 (2014). Issue ISSN 2277-8616 369. <http://www.ijstr.org>.

[23] International Telecommunication Union, Green Power for Mobile Networks. <http://www.itu.int/itunews/manager/display.asp?lang=en&year=2009&issue=04&ipage=32&ext=html>, 2015.

[24] T. Jenkins, A Solar Choice for Pumping Water in New Mexico for Livestock and Agriculture. New Mexico State University's (NMSU) Department of Engineering Technology. <http://www.nmas.org/Journal-v46/V46-p217-Jenkins.pdf>.

[25] A. Kankudti, NTT DoCoMo Tests "Green" Base Stations with Dual Power Source technology, Nex Telecom Asia. <http://nextelecomasia.com/passive-infrastructure/ntt-docomo-tests-green-base-stations-with-dual-power-source-technology>, 2015.

[26] Livestock Watering. Arkansas Economic Development Commission Energy Office. <http://arkansasenergy.org/solar-wind-bioenergy/solar/information-and-resources/livestock-watering.asp>.

[27] M. Morris, V. Lynne, Solar- Powered Livestock Watering Systems. NCAT Agricultural Energy Specialists. Freeze Protection for Livestock Watering Systems, 2002.

[28] M. Morris, V. Lynn, Revised by Lee Rinehart, Solar-Powered Livestock Watering Systems. The University of Tennessee Agricultural Extension Service. PB 1640. NCAT Agriculture Specialists. <https://attra.ncat.org/attra-pub/viewhtml.php?id=251>, 2010.

[29] D. Markham, Solar-Powered Cell Towers Bring Off-Grid Energy to Remote Communities. Technology/Clean Technology. <http://www.treehugger.com/clean-technology/solar-powered-cell-towers-bring-off-grid-energy-remote- communities.html>, 2013.

[30] C. Olsson, H. Dalarna, Cost Analysis on Solar-Powered Radio Base Station with Cooling Demand. by Master's Level Thesis. <http://www.researchgate.net/publication/29752607_Cost_Analysis_on_Solar_Powered_Radio_Base_Station_with_Cooling_Demand>, 2002.

[31] G. Pande, A Case Study of Solar-Powered Cellular Base Stations, University of Gavle. <http://www.diva-portal.org/smash/get/diva2:234252/FULLTEXT01.pdf>, 2009.

[32] Providing Clean, Safe Water Where It's Needed Most. GPXP 10,000 Mobile Trailer Ultra-Filtration Mobile Trailer. <http://www.genproenergy.com/genpro-products-solutions/product-catalog/water/water-filtration-and-purification/fresh-water-purification/gpxp-10-000-mobile-trailer.html>.

[33] S. Qazi, F.A. Qazi, Green technology for disaster relief and remote areas, in 121st ASEE Annual Conference & Exposition, Indianapolis, IN, 2014.

[34] Renewable Energy to Power 4.5% of Mobile Base Stations by 2014. Retrieved from: <http://www.cellularnews.com>.

[35] D.L. Roberts, Safe Drinking Water When & Where It's Needed Most. <http://cleantechnica.com/2014/04/15/safe-drinking-water-needed-2>, 2014.

[36] Reverse Osmosis Layer Membrane. <http://www.h2odistributors.com/reversemembranes.asp?gclid=CI3968D9iLICFUXf4Aodm3QAwQ>.

[37] W. Shiyue, Power Remote Base Stations with Solar Energy. <http://wwwen.zte.com.cn/endata/magazine/ztetechnologies/2009year/no7/articles/200907/t20090710_173704.html>, 2010.

[38] Solar Water Pumping. <http://www.solarray.com/TechGuides/SolarPumping_T.php>, 2015.

[39] S.K. Sahu, M.G. Schultz, G. Beig, Critical Pollutant Emissions from the Indian Telecom Network. Published by Elsevier Ltd. Atmospheric Environment 103. <http://www.elsevier.com/locate/atmosenv>, 2015.

[40] Solar Power in Cellular Base Stations Technical Whitepaper SolarConnect™ Energy Management Solution for Base Stations. Solar Semiconductor. Retrieved from: <http://www.capacitymagazine.com/Article/2788162/Solar-power-in-cellular-base-stations.html>.

[41] B.J.A. Tarboush, D. Rana, T. Matsuura, H.A. Arafat, R.M. Narbaitz, Preparation of thin-film-composite polyamide membranes for desalination using novel hydrophilic surface modifying macromolecules, J. Membr. Sci. 325 (2008) 166–175.

[42] D. Talbot, Solar-Powered Base Stations Can Link Up Remote Areas‖, Retrieved from: <http://www.technologyreview.com/news/417442/a-50watt-cellular-network/>, 2013.

[43] M.G. Thomas, Water Pumping: The Solar Alternative, Photovoltaic Systems Design Assistance Center Sandia National Laboratories Albuquerque, NM 87185-0753. <http://prod.sandia.gov/techlib/access-control.cgi/1987/870804.pdf>.

[44] K. Tweed, Why cellular towers in developing nations are making the move to solar power renewable energy is beginning to replace diesel in cell-phone networks, Sci. Am., 2013.

[45] T.C. Rolla, Sun and water: an overview of solar water treatment devices. J. Environ. Health. <http://solarcooking.org/sunandwater.htm>, 1998.

[46] R. Van Pelt, Solar-Powered Groundwater Pumping Systems for Domestic Use in Developing Countries, Colorado State University Extension, Fort Collins, CO, 2007.

[47] J. Yago, Build your own solar-powered water pumping station, Backwoods Home Magazine Issue 91. <http://www.backwoodshome.com/articles2/yago91.html>, 2005.

Chapter 6

PV Systems Affordability, Community Solar, and Solar Microgrids

6.1 INTRODUCTION

The global solar market has soared as a result of declining prices of solar energy, national incentives, and favorable state policies for solar power throughout the world. It is predicted that installation of solar systems could be tripled by 2017 as compared to the current market. These factors have resulted in an expansion of solar deployment in the middle- and upper-income households but making it unattainable for low-income households on a large scale. According to the 2010 US Census data, 146 million people (or 48% of the US population) are considered low income or poor. While the cost of solar energy has decreased in recent years, many of the world's 1.3 billion off-grid population still find the price tag on most solar products expensive. Since it is expensive to install PV systems, an increasing trend in the developing countries is the funding of locally-based enterprises that can provide these systems in an affordable way through microfinancing. In some remote areas of the developing countries, the end users have no access to banking and financial services. As a result, such services are provided by microfinancing institutions (MFIs). In other developing countries, solar companies have coupled their solar products with financing plans based on the customer's income. Under these plans, customers' payments are scheduled on a monthly or weekly basis for a period of 3–18 months. Other companies in developing countries sell solar energy as a service similar to the way people pay their electric bills in developed countries. In this case, each customer has to pay to keep their services.

In the United States, low-income households spend more than twice (15–20 % of their income) on energy bills than middle- and high-income households. As a result, low-income people have four times the median national burden on their incomes from the cost of energy which puts a strain on their already tight budget. New community partnerships are emerging between citizens, utilities, and governments to provide electricity at a reduced rate for

Standalone Photovoltaic (PV) Systems for Disaster Relief and Remote Areas.
DOI: http://dx.doi.org/10.1016/B978-0-12-803022-6.00006-X

177

low-income people. This is in addition to the sponsoring of several federal and state initiatives to finance solar PV systems and newly-created programs for home-based energy from solar companies. Such initiatives include community solar and solar microgrid; these are important because of the increasing vulnerability of the electric grid from weather-related disasters. Community solar or solar gardens allow low-income people to "own" a portion of the energy from a solar electric generating facility operated by an electric utility or other entity. The PV generated electricity is credited towards their electricity bill similar to what would happen if low-income people installed solar panels on their roof or property. A solar microgrid is an autonomous system that can operate independently and can also be connected to the grid during power outages. It provides backup power even if the grid is not restored. The usage of solar photovoltaics, fuel cells, and windmills in community microgrids as alternative energy producers has increased tremendously since Hurricane Sandy in 2012 caused 8.5 million people in United States to lose power. The goal of community solar and solar microgrids is to make PV generated electricity affordable for low-income people and the electric grid more "resilient" by providing backup power to keep the lights on when a weather-related disaster brings down the grid.

6.2 AFFORDABLE PV SYSTEMS PROGRAMS IN UNITED STATES

Increases in federal and state tax incentive, advances in solar energy, and new and creative financing models have made solar projects, including community solar projects, more financially feasible. Many states in the United States have taken steps to transform the market where solar is available to all residents by adopting new policies, creative incentives, education, and financing solutions. The example of such a program is the case of California State and the District of Columbia (Washington, DC) that are working to provide electricity through solar energy to low-income residents. The State of California started this market transformation program in 2006 and its success is shaping the initial strategies of a relatively new solar program in the District of Columbia that started in 2012. Many pilot programs and creative incentives have also been initiated in the United States and worldwide to make solar accessible for low-income and remote areas consumers. One such program among many in United States is GRID Alternatives that have been bringing solar energy to low-income families in California, Colorado, New Jersey and New York since 2004.

In order to make solar power systems affordable, the Department of Energy have published a guide, a federal overview, for financing solar energy systems for lenders and customers. A partial list of some of these programs with details is below:

- California's innovative single-family affordable solar homes (SASH) program
- District of Columbia's renewable energy rebate program

- GRID Alternatives
- US Department of Energy borrowers guide for financing solar energy systems

6.2.1   California's Innovative SASH Program

SASH was started in 2006 with a 10 year, US$2.2 billion incentive created by the California State Legislature to create a transformative and sustainable solar market in the state of California. Ten percent of California Solar Initiative (CSI) budget amounting to US$216 million was set aside by the legislature. Under CSI, two low-income programs, namely the Multi-family Affordable Solar Housing (MASH) program and the SASH program, were established by the California Public Utilities Commission. During the time of launching the SASH program in 2009, much of the efforts were directed toward supporting low-income communities for the long term. The program also ensured that the infrastructure was developed to support significant growth in the later years. The SASH program installed fewer than 100 PV systems in 2009, over 250 in 2010, 800 in 2011 and over 1200 solar systems in 2012.

6.2.2   District of Columbia's Renewable Energy Rebate Program

The District of Columbia Sustainable Energy Utility (DCSEU) started a program for low-income single-family homes in January 2012 under the Clean and Affordable Energy Act of 2008 by the council of the District of Columbia. The program was designed to reduce energy consumption and create awareness of renewable energy in Washington, DC. It was also designed to form a contract with the Department of the Environment to achieve social equity, economic development, and job creation. The initial goal of the DCSEU was to install solar systems on 20 homes belonging to low-income residents of Wards 7 and 8 in the DC neighborhood. This area is considered low income since household residents spend nearly twice of their total earning on energy bills as compared to the average US household. The goal of installing solar systems in 20 homes was completed in Nov. 2012 and was expanded to include 67 qualified participants. In 2012, the DCSEU program saved enough energy to power over 2000 homes for 1 year and served over 18,500 District households with energy efficiency and renewable energy initiatives. Each participant in this program benefited from no-cost electricity and saved approximately $350–$500 each year. Additionally, the roofs of their homes required less maintenance due to protection from the installation of solar panels on the roofs.

6.2.3   GRID Alternatives

GRID Alternatives have been bringing solar energy to low-income families in California, Colorado, New Jersey, and New York since 2004. This organization has saved nearly US$120 million in lifetime energy costs for 5000 families in California by charging the homeowner only 2 cents for every kilowatt-hour that the solar panel produces. Such a system results in 80% savings in energy bills and reduction of 340,000 tons of greenhouse gases over the system's lifetime. Grid Alternatives is a nonprofit organization that brings together community partners, volunteers, and job trainees to implement solar power and energy efficiency for low-income families, providing energy cost savings. It also provides valuable hands-on experience, and a source of clean, local energy that benefits all. The organization so far has trained more than 15,000 volunteers and job trainees that have received hands-on solar installation experience. Grid Alternatives has 10 regional offices and affiliates serving United States and staff in Nicaragua.

Additional to GRID Alternatives, there are other nonprofit organizations in United States that help low-income individuals use solar energy or gain job training and practical experience:

- Citizen Energy (http://www.citizen-energy.com)
- Community Power Network (http://communitypowernetwork.com)
- Evergreen Cooperative (http://evergreencooperatives.com)
- Nebraskans for Solar Energy (http://www.nebraskansforsolar.org)
- New Vision Renewable Energy (http://www.nvre.org)
- Renewable Energy and Electric Vehicle Association (REEVA) (http://communitypowernetwork.com/node/9371)
- Plymouth Area Renewable Energy Initiative (http://www.plymouthenergy. org)
- Solar Richmond (http://solarrichmond.net)

6.2.4 US Department of Energy Borrowers Guide for Financing Solar Energy Systems

Although energy from the sun is free, the equipment needed to convert it to electricity or heat for buildings can be costly and unaffordable for the common household. In order to make solar power systems affordable, the Department of Energy (DOE) published a guide, a federal overview, in 1998 for financing solar energy systems for lenders and customers. The second edition of the *Borrower's Guide to Financing Solar Energy Systems* which includes both solar electric (photovoltaic) and solar thermal systems was published on Mar. 1999. The guide includes financing resources from Fannie Mae, the Federal Home Mortgage Loan Corporation (Freddie Mac), the US Department of Agriculture, Energy, Housing & Urban Development, Veteran Affairs, US Environmental Protection Agency and US Small Businesses. This guide also includes information about other ways to make solar energy systems more affordable, as well as descriptions of special mortgage programs for energy-efficient homes. This publication was produced by the National Renewable Energy Laboratory, which is a national laboratory for the US Department of Energy Office of Energy Efficiency and Renewable Energy.

In 2013, DOE published another *Guide to Federal Financing for Energy Efficiency and Clean Energy* which provides a snapshot of federal resources that support projects and companies involved. These companies create jobs, spur private investment and invigorate local communities. The DOE sponsors an online guide called the *Database for State Incentive for Renewable Energy and Energy Efficiency* (DSIRE) which describes various tax credits and state-specific incentives for investment in building upgrades and renewable energy projects. This guide covers multiple agencies and specific programs in all 50 US states. Business owners, homeowners, investors, policymakers, and others can use this guide as a "Yellow Pages" for finding federal financing resources.

6.3 AFFORDABLE PV SYSTEMS PROGRAMS FOR DEVELOPING COUNTRIES

According to a recent report, over a quarter of the world's population have no access to electricity and about 2 billion people are forced to use biomass as a fuel for cooking and heating. This can be harmful to human health and bad for the environment. One viable solution is to provide clean electricity through solar energy, which has become less costly in recent years. Although it is still not affordable for low-income populations in developing countries, many corporations have started to manufacture and distribute solar products. These products include solar flashlights, cell phone chargers, radios, small TVs, and solar home systems (SHS) that can power multiple lights. As a result of this trend and new ongoing technological development, the efforts of developing countries to widen their access to clean energy depends less on technology and more on financing arrangements. The lack of suitable end-user finance schemes backed by a policy environment has impeded the low-income population from getting cheap solar energy. Realizing the importance of using clean energy for national development and the concern for the environment, the United Nations General Assembly designated 2012 as the international year of sustainable energy for everyone to recognize the crucial role of energy in sustainable economics. Many national and international organizations have been initiating new programs and incentives to address this issue. A list of some of these programs with details is given below:

- CleanStart
- Renewable energy microfinance and microenterprise program (REMMP)
  - REMMP partnership in India
  - REMMP partnership in Haiti
  - REMMP partnership in Uganda

6.3.1 CleanStart

UN Secretary-General Ban Ki-moon is leading a new global initiative called "Sustainable Energy for All." The goal of this initiative is to meet three objectives by 2030: ensuring universal access to modern energy service, doubling the rate of improvement in energy efficiency, and doubling the share of renewable energy in the global energy mix. To meet these objectives, the UN Capital Development Fund (UNCDF) has created a US$26 million global program called CleanStart in partnership with the United Nations Development Program (UNDP). The goal of this program is to increase access of sustainable clean energy to low-income families through microfinance services. These services are supported by an enabling policy environment and energy value chain.

CleanStart targets low-income clients through microfinance institutions and promotes suitable financing arrangements coupled with support for quality assurance measured for end users. The program also addresses key gaps in energy value chains to contribute to a mutually beneficial cycle of investment and building awareness. This program also creates a new market segment for participation institutions with higher returns. The CleanStart program will be implemented in six African and South Asian countries, and aims to help 2.5 million people move out of energy poverty by 2017. Upon successful implementation, the program will be used to create a business model to be replicated in other developing countries.

6.3.2 Renewable Energy Microfinance and Microenterprise Program

In 2010, USAID started a new program, the Renewable Energy Microfinance and Microenterprise Program (REMMP) in partnership with Arc Finance to test a range of business models. These models are designed to increase financial access of low-income people to clean energy around the world. The goal of REMMP is to increase end-user access to finance by demonstrating the commercial viability of a range of consumer payment models (including microfinance, crowd-funding, remittances and pay-as-you-go models) and by facilitating investment for clean energy finance. USAID is currently working to provide power to India, Haiti and Uganda through its 4-year grant of US$5.6 million to the REMMP program. The goal of the program is to expand the availability of consumer financing for clean energy products which will help low-income people buy devices that improve their income and quality of life while helping USAID's partners to reduce carbon emission. This is achieved by focusing on household and community-scale renewable technologies for communities with low electrification rates. Partnering companies manufacture and sell these products to institutions that finance small scale renewables. Most of the developing countries' renewable energy enterprises are limited in scale due to a lack of end-user finance and distribution. The REMMP approach for scale-up is built on understanding of the relationship between finance, distribution, and technology. This tailored approach takes into account the market conditions, constraints and levering of existing financing, and distribution channels for easier and quicker growth. The REMMP and its implementing partner, Arc Finance, support the companies that manufacture, sell and finance small renewable energy devices such as clean stoves, solar lanterns, and other items used by low-income families. Arc Finance, established in 2008, is a US based global nonprofit company that brings together practitioners, funders, pro-poor enterprises, and end users to develop solution for access to finance for clean energy and water. Arc Finance is currently working with 13 different countries to develop and test business models such as crowd funding, remittance and traditional microfinancing.

Arc Finance has established partnerships with several key organizations throughout the world with specialized technical and local expertise to carry out the goals of REMMP. The details of these not-for profit companies in India, Haiti, and Uganda implementing REMMP programs are given below.

6.3.3   REMMP Partnership in India

REMMP has established partnerships with the following organizations in India.

- Bandhan Knnagar is a part of Bandhan Financial Services Ltd. This company provides microfinance services to 4.8 million rural women through 1864 branches across 19 Indian states. Arc Finance is supporting to develop and launch a new multistate, energy focused subsidiary providing both sales and credit for end clients.
- Friends of Women's World Banking-India (FWWB-I) is an Apex (Asian Professional Exchange) organization that provides support and program development assistance to a network that includes 25 microfinance and community-based organizations located throughout India. Arc Finance is partnering with FWWB-I to create and grow a revolving credit facility for clean energy. Arc Finance through REMMP is providing technical assistance to the sub-partner of FWWB-I's microfinancing institute for a range of energy services.
- Indian Grameen Services (IGS) is an NGO that serves as the action research, pilot testing, and incubation arm of the BASIX Group, which is one of India's largest financial and business development institution. These services focuses on rural poor households and urban slum dwellers. Arc Finance is partnering with IGS to launch energy loan (Urja Samruddhi), and will market, sell, and finance energy products for low-income consumers in rural communities. Training and supporting a network of village-based solar retailers–technicians will also be undertaken as a part of partnership.
- Milaap is an online-funded company that supports the growth of microcredit products such as energy, education, sanitation and water loans by providing inexpensive flexible funds to Indian MFIs through crowdsourcing. The goal

of Arc Finance partnering is to establish low-cost revolving credit facility for energy lending Indian MFIs to simulate their interest in energy sector and reduce borrowing cost of their customers. In addition Arc Finance is providing technical assistance to Millap's subpartners to finance a range of energy services such as solar, clean cooking stoves and microgrids.

- Simpa Networks is a venture-backed energy company that markets high quality SHS with a mission to make modern energy affordable, simple, and accessible for everyone. The SHS are sold on progressive purchase basis which enables off-grid consumers to pay for solar services in a flexible increments over time similar to a pay-as-you-go model. The consumer makes a series of payments under this model, which makes a solar home system available for a paid amount of energy consumption. Arc Finance is partnering with Simpa Networks to test one of its B2C (business to customers) pay-as-you-go business model.

6.3.4 REMMP Partnership in Haiti

REMMP formed a partnership with SogeXpress in 2012 to improve Haitians' access to solar energy through the Arc Finance project. SogeXpress is one of the subsidiary of the leading Sogenbank group which specializes in transfer payments or foreign remittance of Haitians living abroad. Arc Finance in partnership with SogeXpress are working on the remittance-based business model which allows Haitians in the Diaspora to send solar products to their families in Haiti in lieu of cash transfers. The project promoted a remittances-based business model in which the sender uses remittances to buy the energy product from a money transfer organization that has agreed to provide a range of solar products, including cell phone charging systems, to receivers in Haiti. Haiti received nearly 2 billion in remittance amounting to a quarter of the country's GDP.

To accomplish this, SogeXpress has been working to encourage remittance customer to ship solar lamps to their recipients though 12Tel (a Florida based pre-paid communication service company) agent in Florida. Solar lamps carefully chosen for their quality and durability are sent from 12Tel Miami transfer offices to 57 SogeXpress and 57 service centers across Haiti. These portable solar lamps are easy to use; they produce bright light and are equipped with cell phone chargers that are guaranteed for 1 year at an affordable cost. In 1 year, from Apr. 2012 to Apr. 2013, 6136 solar lamps had been remitted and nearly 30,680 Haitians have benefitted from reliable and affordable energy systems. Nearly 9500 children can now study at home by the solar light instead of using kerosene lamp or candles which are bad for human health and environment. SogeXpress and Arc Finance are playing a pioneering role in enabling access to solar energy through this project as only 12.5% of Haitians have access to electric grid. After a successful year, SogeXpress created incentives for its agents to sell lamps in the streets and 2000 solar lamps had been sold by 500 agents within 2 months.

6.3.5   REMMP Partnership in Uganda

REMMP partnership in Uganda is through SolarNow which was established in 2011 to sell SHS in Haiti. More than 95% of Uganda's rural population have no access to the electrical grid. The SHS costing between US$475 to US$1400 can light several rooms and energise other household items such as radios, cell phone and TVs. SolarNow is a private solar distribution and service company that offers credit to customers on a hire purchase basis using an inhouse-created "PayPlan" business model. Such a plan enables the customer to purchase the solar home system on a monthly installment for a period of 1 year until they have covered the full price, at which point ownership is transferred to the customers. Arc Finance under REMMP will help SolarNow attract financing to test the Payplan model, scale operations, and expand to other countries in sub-Saharan Africa on a sustainable long term basis. The customers receive a 2 year service warranty after acquiring the ownership. In SolarNow's hire purchase scheme the customer can double their system's power easily in the payment plan from 50 to 100 W or add an appliance such as a refrigerator or TV. The customers can also sign for another 12 month payment agreement to cover the cost of upgrade for which the payments will be lower than those in the initial 12 months. This feature of hire purchase agreement has helped SolarNow in on-time payment as well as reassuring the customer with an established payment record with lower-cost underwriting and reduced risk in the second time payment agreement. SolarNow, as of Feb. 2014, are providing solar power to 3819 Ugandan customers.

6.3.6   Companies and Organization Bringing Solar Energy to Developing Countries

Clean Tech energy has listed 40 companies and organizations in the developing countries that are making tremendous differences by helping low-income people in the world to gain access to electricity, cleaning up the climate by avoiding the use of diesel generators, kerosene lighting, and burning of coal which is also harmful for human health. These organizations are also saving lives by providing cheap electricity and creating tens of thousands jobs mainly in the remote areas. Bennu Solar, aChina-based reliable chain (http://bennu-solar.com/resources/) has listed more solar companies, both not-for-profit and private enterprises by dividing developing world in four regions: Developing Asia, Africa, Latin America, and the Middle East. The list includes names of countries in each region, name of solar companies and details of consulting regarding the supply of solar technology to the rural poor with procurement support in developing countries.

In addition, the following companies (partial list only) are working to assist people living in energy poverty regions of the world by providing solar energy solutions for their educational, health, agricultural, and economic development:

- Solar electric light fund: Energy is human right (http://self.org/)
- International Renewable Energy Agency (IRENA)
  (http://www.irena.org/menu/index.aspx?mnu=cat&PriMenuID=13&CatID=9)

- Sunfunder (http://sunfunder.com/about/sunfunder)
- SunnyMoney: A brighter solution (http://sunnymoney.org/)
- WISIONS of Sustainability (http://www.wisions.net/)
- Solar Energy International (http://www.solarenergy.org/)
- New Vision Renewable Energy (http://www.nvre.org/)
- Partners in Health: Health is human right (http://www.pih.org/)
- Sun Energy Power Corporation: Sustainable Social Entrepreneurship
  http://www.sunenergypower.com/mission.asp

In order to find innovative solutions to create a new and sustainable energy future, Abu Dhabi has established a US$4 million Zayed Future Energy Prize to recognize and reward the innovators. The award was announced in the 2008 World Future Summit to honor the vision of the founding father of the United Arab Emirates, Sheikh Zayed bin Sultan Al Nahyan, for his legacy of environmental stewardship. The prize is awarded annually to five high schools from five world regions, a lifetime achievement recipient, nonprofit organizations, a small, medium or large enterprise/corporations who have made significant contributions to the global response to the future of energy.

6.4 COMMUNITY SOLAR FOR AFFORDABILITY AND RESILIENCE

When disaster strikes, the whole infrastructure including electricity shuts down for days, months, or even years depending on the nature of disaster. In the absence of electricity all human activities are decreased, businesses are either damaged or leave the area. When businesses fail to revive or leave after a disaster, the overall economic health and social viability of a community is threatened requiring measures taken to return the community to normal after providing relief in the early stages of disaster. Community solar is a partnership between citizen, government and businesses. It is based on a PV electric system that provides power and/or financial benefits to the participating members of the community through a voluntary program. In the case of utility-sponsored programs, the customers can either buy or lease solar panels in a centralized solar array and receive regular credits via their utility bills. The "Resilient Community" Partnership is a cooperative framework that is essential to fostering community disaster resilience. The goal of this partnership is to maintain the economic and social viability of the community following a disaster. Businesses are key factors in bringing the community return to normal which provide jobs that generate salaries, in turn driving consumption and results in generating taxes that support the governments' ability to function. The benefits of community solar are given below:

- No upfront investment. There are no upfront costs to install solar panels as it is the collective community project
- No maintenance or repair costs. There are no repair costs or panel maintenance in a community solar project

- Reduced greenhouse gas emissions. It is environmentally friendly as it uses renewable sources of energy to produce power
- Suitable for renters, condo owners and homes who would otherwise not be suitable for a rooftop system
- Cancel at any time. It is easy to cancel any time after the contract period limit
- Fixed price for certain number of years
- Increased solar production. Community Solar panels usually track the sun throughout the day to produce higher energy than a stationary rooftop system
- Moves with you. Some community solar program unlike solar rooftop panels can easily move with you to another residence

6.4.1   Community Solar in United States

There are over 20 community solar projects in United States and continue to increase in numbers. According to Solar Garden Institute, the idea of community solar in United States was first conceived in 2003 by the City of Ellensburg, Washington State University Energy Extension, and the Bonneville Environmental Foundation. In Germany and Denmark the idea of community ownership of renewable energy started much earlier and goes back to owning shares in a wind turbine or solar array. The term "solar garden" was first used to describe utility scale facilities in United States and Spain. In United States the first Community Solar Garden was started by Luke Hinkle of My Generation Energy Inc., who constructed and maintained the Brewster Community Solar Garden Project in 2009.

The pioneering work by the City of Ellensburg projects was completed in 2006 by the financial support of businesses, local homeowners, and Central Washington University where the solar panels were leased to utility customers. The project has grown to 78 kW from 36 kW in its third phase and has since produced more than 170,000 kWh, averaging 58,000 kWh annually. In 2007, the Sacramento Municipal Utilities District (SMUD) started the nation's largest solar garden, 1 MW in size, where the subscribers enter into a power purchase agreement with the utility. The Clean Energy Collective in Colorado has built a 78 kW solar array on the purchase of 20 shares by local community members.

In Colorado, the first community solar facility was built in 2008–09 at Brighton by United Power Sol partners. As part of the program, 48 people purchased (leased) a 210-W solar panel for a 25-year period. The customer's benefits include solar credits on their electricity bills which will produce nearly a three-percent return on its investment or approximately $32 per year in electricity credits. In 2010, the Colorado House Transportation and the Energy Committee approved a bill that would extend the rebates and renewable energy credits to homeowners when they install solar energy systems on their property when solar facilities are jointly owned by 10 or more customers at a shared location. The first community-owned solar garden in Colorado started in Aug. 2010 at El Jebel. The 340-panel solar garden is built on unusable land in the Roaring Fork Valley and was developed by clean energy collective. It is a 78 kW solar array purchased by 20 members of local community. The system was connected to grid with partnership by Holy Cross Energy local electric cooperative which collects the power produced by the solar garden and then directly credits owners' utility bill each month.

In Maryland, a group of individuals in University Park installed a 22 kW rooftop solar array in 2010 on a local church by establishing a limited liability corporation with 30 community members investing. The 81 solar panels are installed at the Church of the Brethren, which will generate 230 W each day or about 28,000 kWh/year.

6.4.2 Making a Solar Garden

One way of making community solar is through a solar garden: solar systems that are community-owned and shared. Instead of installing solar panels on individual's roofs, large numbers of solar panels are installed in a central location within a town or city limit to expand the availability of solar energy. The goal is to allow renters, or anyone else who does not or cannot put solar on their rooftops, but still wants to benefit from localized solar electricity generation. In some countries it is encouraged by federal and state governments through rebate and tax incentive programs. Electricity from the solar panels can be stored in the batteries for supplying to hospitals, schools, military bases and other needs in the aftermath of disasters. It can also be sent to the grid where it is sold to the local utility, which then credits the sale to the owners or subscribers of the solar garden.

The subscribers may purchase a portion of the power produced by the solar panels and receive a credit on their electric bill, just the same as if the panels were on their own roof, using virtual net metering. The customers within the solar garden's service area, including businesses, residents, nonprofits, local governments, and faith-based organizations, can all subscribe to the solar garden. A solar garden is a distributed generation project and provides benefits to communities by affordable energy, creating local usage and avoids destroying delicate habitats. It also bypasses the need for inefficient transmission lines, which lose power during transmission and can take many years to put in place. The solar garden helps the community save money by pooling resources and buying panels as a group, and give subscribers a lower cost than going it alone. A schematic of solar garden is given in Fig. 6.1.

The solar gardens can be located on a large roof, in a parking lot, unused field or commercial area usually within a town or city limit. The size of a solar garden depends upon a country, state or the area in which it is located. Each country or state allows its residents to make solar gardens based on the need and newly-passed government bills. The size of solar gardens in the state of

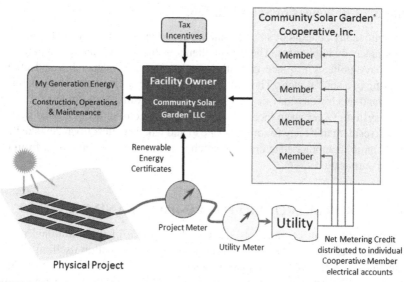

FIGURE 6.1   Community Solar Garden. *Graphic courtesy of My Generation Energy, Inc. Community Solar Garden is a registered trademark of Brewster Community Solar Garden, LLC.*

Colorado, United States, ranges between a large roof to 16 acres, accommodating 10 kW to 2 MW. In the state of California the size of solar garden could be 160 acres accommodating up to 20 MW. The Solar Garden Institute of United States (Solar Garden Institute http://www.solargardens.org/) has outlined the following steps to organize a solar community garden. The organizer needs to work concurrently in three different areas: policy, community organization, and project development.

- Work on policy to promote community power for support of solar gardens. This can be achieved by working with your local utility, legislators, and county/area planning commission to develop solar gardens programs and zoning rules. Be prepared to support and work toward nationwide policy dealing with solar gardens.
- Organize communities by arranging meetings, partnering with local non-profit and recruiting early adopters. Work with neighborhood associations to find out about parking lots, religious places, and unused lands for asset mapping in the community. This will also help you to find host sites on large roofs and suitable properties near utility distribution lines or substations.
- Arrange the bidding process for the construction of solar garden by identifying local solar companies involved in solar panel hosting.
- Locate and recruit subscribers from businesses, city governments, and non-profit power users, in addition to regular subscribers who can use a significant amount of the power.

6.4.3 Challenges of Solar Gardens

Despite many benefits of solar community gardens it has many challenges and hindrances to implement. The first challenge is that every state or country does not have policies in place that allow such communal ownership and credit-sharing of a solar system. The next challenge is price of land and its availability. The land in a big city may be very expensive or unavailable to implement a suitable system. Other issue includes the minimum size of a piece of solar system that a resident must pay for or subscribe to and whether subscriber can take their ownership to another state or town which may not have a solar garden from the same corporation.

6.5 SOLAR MICROGRID FOR RESILIENCE AND DISASTERS

A microgrid is an autonomous scaled-down version of the traditional power grid that is able to balance generation and consumption within itself on a much smaller scale. It could be as small as a small disaster-ridden town, an offshore oil rig, or as large as a military base. A minigrid is larger than a microgrid and usually encompasses a single village or a cluster of villages. A solar nanogrid is defined as PV system of 5 kW for remote areas not tied to the grid and 100 kW for grid-tied systems. Applications of nanogrid applications range from providing basic electricity services to people living in poverty to supply emergency power for commercial building. The microgrid might use storage to buffer distributed renewable energy resource like solar PV, or it might simply fire up a fuel-burning generator. The components necessary to provide power include batteries for energy storage, a power electronic converter, software, and hardware. Microgrids can operate independently or in parallel with the traditional power grid. According to a recent research report by Navigant Research, it is predicted that the total worldwide capacity of distributed power generation in microgid will increase by more than five times in the next 6 years growing from 746 MW in 2012 to about 4000 MW in 2018. This increase will generate more than US$12.7 billion in revenue for businesses. It is also projected that nanogrid business will reach up to US$60 billion in revenue by 2023.

Microgrids have the ability of self-insulating, termed "islanding," which will allow the microgrid to generate its own power to supply to the locations affected by weather-related outages and disasters. One example of the microgrid generating its own power during a disaster is the case of New York University's (NYU) self-sufficient microgrid system. This system supplied power to 26 electrically connected buildings in the aftermath of Superstorm Sandy in 2012. The NYU's microgrid is designed to distribute electricity independently of the utility's main grid, which consists of a 13.4 MW combined heat and power plant made up of two giant natural gas-fired turbines housed below Mercer Street in New York. In addition to supplying power to 26 buildings, the microgrid also provides hot and cold water for up to 40 buildings by harnessing heat that would otherwise be wasted. The demonstrated performance of microgrids in the aftermath of superstorm Sandy, has resulted in leaders of the US government in the North Eastern states to identify microgrids, like the NYU, as a key component to improve energy resilience and mitigating the impact of weather related disasters. Other examples include the US Federal Administration's (FDA) White Oak research facility in Maryland, where the microgrid supplied power to all of the FDA building on the campus for two and a half days after the local grid failed to provide power during Superstorm Sandy.

Because of much lower line losses in the transportation of electricity, off-grid institutions like prisons, campuses, military operations, and large commercial and industrial markets in remote settings are building and maintaining their own microgrid. In places like Africa, Brazil, Haiti and India that have never had access to reliable grid power, microgrids are replacing expensive and polluting fuels like diesel and kerosene. In places where grid power is available, micro-grids are being built to justify the high cost or risk of an outage and be self-reliant. Microgrids have served as disaster recovery apparatus in the aftermath of natural disasters (see also Fig. 6.2).

6.5.1   Benefits of a Microgrid

- Less vulnerable to outside cyber or physical attacks because of its ability to shut itself off from the main grid (islanding)
- More efficient as very little energy is lost in transmission because the source of generated electricity is so close to the users

FIGURE 6.2   Solar microgrid.

- More reliable because closer to source, smaller in size, uses power generation to specific places and is able to respond to demand more quickly
- Easier to make expansion and update because the layout is modular
- More specific to the design and future planning of the participating entities because of local control
- Smaller infrastructure because of fewer load sources

6.5.2 Challenges of Developing Solar Microgrids

- The cost of installation of distributed energy resources such as solar and wind may be too great for some areas
- The regulation and policies in many countries may need to be addressed for regulating the operation of microgrids
- Lack of safety and operation standards as well as integration to data on power quality from different sources on microgrids
- In the absence of suitable infrastructure, there is the possibility of market monopoly against pricing abuse
- Lack of technical experience, infrastructure, and communication protocols

Despite these challenges, the concern about utility grids, greater availability of distributed generation, rising costs of fuel and a drop in the cost of photovoltaic systems is driving the microgrid technology into the brighter future.

6.6 CASE STUDY OF SOLAR MICROGRIDS IN INDIA

Most of the solar microgrids in Africa, Asia, or South America are small systems used to power a few homes in rural areas. However, efforts are underway throughout the world to increase the number of these distributive systems to supply electricity in the aftermath of disaster or to provide access to electricity in remote areas. In India, 33% of people in the rural areas do not have access to electricity. Since economic expansion is on the rise because of the growing population, an increase in the demand electricity is the resultant. This increase in demand of electricity and 40% loss in generated electricity due to inefficiencies and theft are placing severe strain on the countries' energy infrastructure. India's annual electricity demand is expected to increase from 900 billion kWh to 1400 billion kWh by Mar. 2017. In order to meet this demand, India's Ministry of New and Renewable Energy (MNRE) in 2010 announced the goal of deploying 20 GW of grid-connected solar power and 2 GW of off-grid solar by 2022 under the Jawaharial Nehru National Solar Mission (JNNSM). The goal of the JNSM phase 1 project was to implement 200 MW to off-grid solar power for the electrification of rural areas and 1 GW grid connected solar by 2013. Nearly all microgrids in India are powered by photovoltaic cells except 20–30 networks that run on hydropower and biomass-powered grids. The purpose of Indian solar microgrids is to bring millions out of darkness by using solar panels and biomass, where 300 million people live without electricity.

6.6.1   Solar Microgrid (Funded by Government of India) for Rural Electrification in the Remote Areas

The Government of India financed a rural electrification project using solar microgrids in 2012 to supply power to 57 tribal villages situated near an ongoing insurgency. It was located in the Visakhapatnam District of the southern Andhra Pradesh state and was completed in 11 months by a local Indian company. The project was implemented by an EPC company, Premier Solar Systems, under the Rajiv Gandhi Grameen Vidyutikaran Yojana program, which promotes rural electricity infrastructure and rural household electrification from renewable and conventional sources. Most of the homes in the villages are located 20–50 km from the main paved road and are only accessed by trails. This made the installation more challenging as material and components were difficult to be transported without roads and thus fewer materials were used (see Fig. 6.3). The generated electricity is used to provide nighttime lighting of streets and houses for 6–8 hours and charging of mobile phones for 1 hour. The solar microgrid acts as a central power supply, which distributes connections to individual houses in the villages. The installation consisted of PV panels, batteries, and a central cabin (power supply center), as well as connections and transmission to the

FIGURE 6.3   Transporting and installing solar panels on foot in remote areas.

individual houses or way stations. Other than some diesel generators, there was no power source for these communities.

6.6.2   System Rating of Indian Solar Microgrid for Rural Electrification

The rural electrification system in India uses a set of remote microgrids ranging from 2 to 13 kW with battery backup for a total output of 365 kWp to supply electricity to 2225 homes. The system is designed to provide each household with 100 W of loads for 6 hours each day whereas each household was supplied with two compact fluorescent lights of 18 W and a power socket for connecting additional loads of approximately 60 W. The power for these loads is made available for 4 hours from 6 pm to 10 pm and for 2 hours staring from 4 am to 6 am. A limited number of street lights are also powered from dusk to dawn.

The base level remote microgrid system consists of 13-kW solar PV with a battery storage serving a 33.2-kWh/day load capacity. The battery bank for the system consists of 1530 Ah, 120 V batteries.

The load represents 57 villages, each with 6 hours of lighting and 2 hours/day of outlet charging. Additional lightning and charging capabilities at a household level covering increased delivery capacity up to 12 hours a day can be obtained by adding more PV panels, wind turbines, diesel generators, and additional batteries.

6.6.3   Benefits of Solar Microgrid in Remote Areas of India

The availability of electricity in the villages gave people a sense of empowerment which affected their lifestyle. People started small businesses and bought TVs and cell phones from their extra income. This made them better informed and connected to the outside world. The children spent more time on their studies and became more competitive in national examinations. The residents used fans in their homes and businesses which helped them to be more productive in a cooler environment. It also cleaned the air, as fewer diesels were used and there were reduced hazards caused by the use of kerosene oil for light and cooking.

6.6.4   Challenges of Solar Microgrid in Remote Areas of India

The completion of this project faced many challenges due to the remoteness of the villages, lack of infrastructure, the cultural issues of the tribal people and ongoing insurgency. These challenges, given below, will be typical to any other area with similar circumstances.

- Lack of trained personnel to complete the project
- Difficult to transport the project material in the absence of roads and takes too long to transport material on foot
- Temporary closing of trails because of insurgency and weather conditions
- Health problems of workers due to lack of clean water and safe food

6.6.5   Solar Microgrid for Rural Electrification in Remote Areas of India (Completed With the Help of a Global NGO)

A solar-powered microgrid for the electrification of a small village, Dharnari, was completed with the help of a global NGO, Greenpeace, in the Indian state of Bihar. It is a battery backup system, shown in Fig. 6.4A. Bihar is considered one of India's poorest states, where 82% of the population is not connected to the traditional grid-based system. Dharnai is located in the district of Jehanabad where most of people are farmers. The installation of 100 kW solar microgrid took only 3 months to complete with the help of Greenpeace. It is operated in association with BASIX, a livelihood promotion institution as well as CEED, a network of NGOs and think-tank organizations in Bihar to support renewable energy development.

The microgrid supplies clean electricity to more than 450 households (2400 people) and 50 commercial establishments. This was achieved by using 70 kW for electricity generation and 30 kW for 10 solar-powered water pumping systems of 3 horsepower each. The electricity is also supplied to health centers, a farmer training center (shown in Fig. 6.4B), two schools, 50 commercial

FIGURE 6.4   (A) The Dharnari solar microgrid; (B) use of Dharnari microgrid for farming.

establishments and 60 street lights. The village have been trying to get connection to the grid for the last 30 years.

6.7 CASE STUDY OF SOLAR MICROGRIDS IN UNITED STATES

Microgrids in United States, once used mostly by colleges and hospitals, are now expanding to military bases, businesses, and local government facilities to avoid the power failures posed by weather-related disasters. There has been a tremendous increase of alternative energy generators powered by solar photovoltaics, fuel cells, and windmills after Hurricane Sandy in 2012, when 8.5 million Americans lost power. These microgrids aim to make the main grid more resilient and provide backup power to keep the lights on when a storm disrupts the grid. Microgrids have evolved from controlling simple generator backup systems into sophisticated smart grids that can ensure reliability, resiliency and energy independence. The microgrids in the United States are much larger than the ones in India. Examples of microgrids in United States include microgrids at the University of California San Diego (USCD) and Nellis Air Force base in Nevada which produces and distributes power to a population of the size of a small city. A relatively small microgrid has recently been installed at Wesleyan University, Connecticut for a campus of 3200 students after widespread outages due to Superstorm Sandy. This microgrid distributes electricity to 312 buildings on the campus without depending on the outside grid and will keep the power on even if the local grid loses power due to weather-related disasters. The State of Connecticut paid US$694,000 of the US$3.5 million cost to establish this microgrid, which is the first state-funded microgrid. Other states in the north east like Massachusetts, Maryland, and New York have launched several initiatives to support the creation of microgrids to improve the resilience of the grid and mitigate the impact of weather related disasters.

6.7.1 Case Study of the UCSD Microgrid

UCSD is a 1200 acre campus, almost the size of a small city with 725 building of 13 million square feet of floor space. The UCSD campus is connected to San Diego Gas and Electric (SDEG) utility grid at a 69 kV single substation. The UCSD microgrid starts at the university substation and includes installed power plants, including a 30 MW cogeneration system containing two gas turbines and a steam turbine, 3 MW solar panels and 2.8 MW fuel cell generating over 42 MW of electricity as shown in Fig. 6.5A, B.

The UCSD currently produces more than 92% of its own energy after starting a state-of-the-art energy-independent microgrid in 2006. The university saves US$800,000 per month on its power bills by producing self-generating energy. The central cogeneration plant (combined heat and power) supplies chilled water cooled overnight when electrical demand is low and is stored in a 3.8 million gallon storage tower. The stored chilled water is used for air

FIGURE 6.5 (A) UCSD microgrid; (B) Solar panels on the roof of USCD buildings. *UC San Diego Jacobs School of Engineering.*

conditioning of campus buildings during the day and the heat removed from the buildings is absorbed by the refrigerant, transferred to the chiller and released to the atmosphere via cooling towers. Most of the energy is generated from the sun using 3 MW of traditional photovoltaic panels, 30 kW concentrating solar and 30 kW of smart solar installation with integrated storage. With 45,000 students, the UCSD campus consumes more than 250,000 MWh of energy annually. The campus controls 42 MW of generation, and then purchases power from the market. The rooftop solar is currently maxed out at about 3 MW of rooftop space with new buildings and parking lots being constructed.

A plant information (PI) system by OSIsoft is installed to create a sustainable environment for learning and innovation by gathering data from hundreds of sensors for synchronizing and coordinating the operation of this complex energy system. OSIsoft is a set of software modules designed for plant-wide monitoring and analysis. The information from the electrical meters are connected to the PI system via the data connections fitted on two rooftop solar arrays allowing the university to know the amount of power the panels are producing at all times.

6.7.2 Master Controller for UCSD Microgrid

The UCSD microgrid is the world's first microgrid with a master controller that monitors and controls the real time operations for optimization. This is achieved by rescheduling power generation, storage, and load based upon dynamic market price signals. It also allows the operators to disconnect the grid from the larger grid to maintain power supply during power outages, and manage the diverse energy generation and storage resources to minimize the cost. The microgrid also contains a large number of synchrophasors, or phasor measurement units (PMUs) as part of cyber-security synchrophasor platform for data collection and processing. The data collected by cyber-security synchrophasor via PMUs includes voltage, frequency and phase which is used both for real time display and for main campus data archiving system. Additional PMUs will be installed on a distribution circuit to understand the performance of energy distribution, photovoltaic penetration and electrical vehicle charging. The current cybersecurity synchronous platform consists of two computers running a standard operating system configured for dual-redundance failover collective system and a standard operating system for visualization, a global positioning system based clock, two switches, and two PMUs. According to North American Electric Reliability Corporation, implantation of synchrophasor systems should be performed securely in redundant pairs to avoid system data gaps while standard maintenance is performed on the system.

The microgrid controller is integrated with OSIsoft's PI data servers on the campus. The data servers are used for managing data collection and data analysis by interfacing with the power analytic controller. The highly instrumented microgrid currently monitors approximately 84,000 data streams per second and is designed for expeditious integration of distributed energy resources. The microgrid controller and the resulting data will be utilized to improve management and efficiencies of the utility's and statewide grid operations. This includes renewable supply, demand response, power outages, load balancing and excess generation performed by the collaborating engagement of San Diego Gas & Electric (SDG&E), California Energy Commission (CEC), the California Independent System Operator (CASIO) and Department of Energy (DOE).

6.7.3 Energy Storage for UCSD Microgrid

Since the sun does not always shine, there is a need to store energy created by the sun so that it can be used when market prices for grid energy are high, or fed back into the grid as a source of revenue. Energy storage is essential to remove intermittency and variability of solar energy and is critical to a resilient, efficient, reliable and cost-effective solar microgrid. In the case of solar microgrid supplying PV-generated electricity to the load, the stable operation of microgrid requires energy storage by managing load supply and supply variations for keeping voltage and frequency constant. The integration of microgrids that include PV-storage systems will require a high level of system control, a detailed knowledge of various loads being served, and thoughtful design of the PV-storage system which in return will provide many operational benefits to customers and utilities. UCSD is currently using and experimenting with the following storage technologies in collaboration with partners from industry including Sanyo, Maxwell, Panasonic and National Renewable Energy Laboratory (NREL):

- Electrochemical batteries: These batteries produce and store electricity by means of a chemical reaction and include lithium–ion batteries, zinc–bromine flow batteries and used electric vehicle batteries. UCSD has developed an experiment to use discarded electric vehicle batteries for stationary energy storage, in collaboration with BMW (car maker) and energy corporation RWE. This is achieved by connecting 330 kW PV systems to worn out batteries with 80% remaining charge capacity for 8–10 years and testing its performance.

    UCSD will also be adding battery storage of 2.5 MW and 5 MWh systems to the existing energy storage. This new energy system will be installed in spring 2015 and is based on high performance lithium–ion iron phosphate batteries. The storage system will be integrated to the microgrid which generates 92% of the electricity used on the campus annually. The batteries can power 2500 homes and are toxic free, nonexplosive and firesafe. This project will receive up to $3.25 million in financial incentives to be completed in 2015 by Self Generation Incentive Program which is a California based rebate program for the installation of clean and efficient distributed generation technologies.

- Ultracapacitor: The principle of using ultracapacitors for energy storage is based upon an electrostatic mechanism. This electrostatic mechanism charge the devices rapidly from an electrical energy source such as, a photovoltaic cell, and discharge their stored energy when needed for consumption. The action of these ultracaapcitors, when connected to a PV system, is to act as a standby reservoir of electrical energy to reduce the variability of solar energy. UCSD uses 30 kW ultra-capacitor-based energy storage systems from Maxwell Technologies, Inc., which will be combined with the existing Soitec's Concentrated Photovoltaic Technology system.

- Thermal energy storage: This energy storage allows excess thermal energy to be collected at one point and use it at a different location at a later time. UCSD uses 3.8 million gallons of thermal energy storage from wasted heat of a plant and use it as a power source for a water chiller. The chiller fills a 4 million gallon storage tank at night with cold water and is used during the warmest time of the day to cool campus buildings.

6.7.4 Challenges of UCSD Microgrid

- Variable energy from the sun makes it unpredictable compared to traditional fossil fuel-based generation
- Difficult to integrate different software platforms for real time control, monitoring and acquisition of data
- Difficult to measure energy accurately because of many buildings of multiple scale and difficult to control at multiple scales
- Higher cost and complex integration

6.7.5 Solar Microgrid With Battery Storage for Emergency Shelters in United States

Another large scale solar microgrid project has begun construction in Rutland, VT, being the first project in United States to establish a microgrid powered solely by solar and battery backup, with no other fuel source. The project will be installed by Great Mountain Power and is partially funded by a federal–state–NGO partnership involving the State of Vermont, the US Department of Energy, Office of Electricity and the Energy Storage Technology Advancement Partnership (ESTAP), which is managed by Clean Energy States Alliance and Sandia National Laboratories. The project is a 2.5 MW project which incorporates 7722 solar panels and has 4 MW of battery storage consisting of both lithium–ion and lead–acid batteries. The energy generated will be integrated into the local grid, and will be used to provide resilient power to critical facilities such as emergency shelters, firehouses and fueling stations during power outages. The solar microgrid will also be used to provide clean, resilient power to an urban community that suffers from frequent power outages due to storms. The project will be located on brownfield land used for burying waste and will be the first "solar + storage" to provide full back up power to an emergency shelter on the distribution network.

BIBLIOGRAPHY

[1] Y. Agarwal, T. Weng, R.K. Gupta, Understanding the Role of Buildings in a Smart Microgrid, Computer Science and Engineering, UC San Diego, La Jolla, CA, 2011.

[2] Arc Finance, (2014). Changing Lives Through Access to Finance for Clean Energy & Water Renewable Energy Microfinance and Microenterprise Program (REMMP). <http://www.arc-finance.org/remmp.html>.

[3] K. Bullis, How solar–powered microgrids could bring power to millions, MIT Technol. Rev. <http://www.technologyreview.com/featuredstory/429529/how->, 2012.

[4] CleanStart, Microfinance opportunities for a clean energy future, U.N. Energy Network. <http://www.un-energy.org/publications/3504-cleanstart-microfinance-opportunities-for-a-clean-energy-future>, 2012.

[5] M. Coren, Solar Microgrids Bring Power to People Who Have Been Off the Grid Forever. <http://www.fastcoexist.com/1679890/solar-microgrids-bring-power-to-people-who-have-been-off-the-grid-forever>, 2012.

[6]   Community Solar Gardens, Clean Energy Resource Team. <http://www.cleanenergyresourceteams.org/solargarden>, Undated.

[7]   CPN Solar Cooperatives & Solar Bulk Purchases, Community Power Networks. <http://www.communitypowernetwork.com/solarco-ops>, 2014.

[8]   Financing Solar Energy Systems, DoE Guide for Lenders and Consumers. <http://www.nrel.gov/docs/fy99osti/26242.pdf>, 2009.

[9]   E. Fitzpatrick, India village claims a first—100% solar, storage micro-grid, RE New Economy. <http://reneweconomy.com.au/2014/india-village-claims-a-first-100-solar-storage-microgrid-81573>, 2014.

[10]   B. Freling, Energy Is Human Right. <http://www.bobfreling.com/solar-ngos>, 2012.

[11]   J. Garthwaite, Solar Micro-Grid Aims to Boost Power and Food in Haiti, National Geographic, 2013.

[12]   International Energy Agency, Energy for All: Financing Access for the Poor—Excerpt of the World Energy Outlook, Paris, France. <http://www.iea.org/media/weowebsite/energydevelopment/weo2011_energy_for_all.pdf> gy, 2011.

[13]   K. Ling, First microgrid reliant on solar to provide backup power for Vt. Town. Renewable Energy, E&E News PM. <http://www.cleanegroup.org/assets/Uploads/eenews-kling.pdf>, 2014.

[14]   C. Marnay, N. Zhou, M. Qu, J. Romankiewicz, Lessons Learned from Microgrid Demonstrations worldwide. China Energy Group Environmental Energy Technologies Division Lawrence Berkeley National Laboratory, U.S. Department of Energy under Contract No. DE-AC02, 2012.

[15]   L. Margoni, Environmentally-Friendly Battery Energy Storage System to Be Installed at UC San Diego, UC San Diego UC San Diego News Center. <http://ucsdnews.ucsd.edu/archives/author/Laura%20Margoni>, 2014.

[16]   J.J. MacWilliams, Guide to Federal Financing for Energy Efficiency and Clean Energy Deployment. <http://energy.gov/sites/prod/files/2014/09/f18/Federal%20Financing%20Guide%2009%2018%2014.pdf>, 2014.

[17]   V. Modi, S. Susan McDade, J. Saghir, The Energy Challenge for Achieving the Millennium Development Goals. New York, NY. <http://www.unmillenniumproject.org/documents/MP_Energy_Low_Res.pdf>, 2005.

[18]   Navigant Research, Microgrids. <http://www.navigantresearch.com/research/microgrids>, 2014.

[19]   C. Nawilis, End-User Financing That Makes Sense: Things You Didn't Know About Solar in Developing Countries. <http://www.motherearthnews.com/renewable-energy/3-things-about-solar-in-developing-countries-zbcz1406.aspx#ixzz3O7xUKc3C>, 2014.

[20]   N. Neha Khator, R. Kumar, Power from the Sun: A New Life for Dharnai, India, Greenpeace International. <http://www.greenpeace.org/international/en/news/Blogs/makingwaves/Dharnai-Live/blog/49962/>, 2014.

[21]   A. Pathanjali, Solar Energy Microgrid Powers India. <http://greenpeaceblogs.org/2014/07/17/solar-energy-microgrid-powers-india-village-bihar/>, 2014.

[22]   B. Paulos, UC San Diego Is Building the "Motel 6" of Microgrids. <http://www.greentechmedia.com/articles/read/byrom-washom-master-of-the-microgrid>, 2014.

[23]   P. Prabakar Loka, S. Moola, K. Polsani, S. Reddy, S. Shannon Fulton, A. Skumanich, A case study for micro-grid PV: lessons learned from a rural electrification project in India, Prog. Photovoltaics: Res. Appl. 22 (7) (2014). Article first published online: 25 NOV 2013. Available from: http://dx.doi.org/10.1002/pip.2429.

[24]   J. Pyper, ClimateWire, Are microgrids the answer to city-disrupting disasters? Sci. Am. <http://www.scientificamerican.com/article/are-microgrids-the-answer-to-city-disrupting-disasters>, 2013.

[25] S. Qazi, F. Qazi, Green technology for disaster relief and remote areas, in 121 ASEE Annual Conference & Exposition, Indianapolis, IN, 2014.

[26] J. Runyon, US State Breaks Ground on a "Perfect" Solar + Storage Microgrid that Can Provide Resilient Power. Renewable Energy World.com. <http://www.renewableenergy-world.com/rea/news/article/2014/08/us-state-launches-solar-storage->, 2014.

[27] K. Reylonds, Brewster Community Solar Garden Cooperative, Inc. <http://www.brewster-communitysolargarden.com/faq/>, 2014.

[28] Solar Garden Institute. <http://www.solargardens.org/>, Undated.

[29] Solar Calfinder, The Country's First Community—Owned Solar Garden, Clean Techies. <http://blog.cleantechies.com/2010/08/18/first-community-owned-solar-garden/#sthash.o3ViQxUG.dpuf>, 2010.

[30] Solar Gardens Community Power, Community Energy Directory. <http://blog.solargardens.org/2010/06/community-energy-directory.html>, 2010.

[31] SRP (Salt River Project) Community Solar, Benefit of Community Solar. <http://www.srpnet.com/environment/communitysolar/benefits.aspx>, Undated.

[32] Z. Shahan, Pioneering Commercial Solar Microgrid Completed, CleanTechnica. <http://cleantechnica.com/2013/12/21/vestas-announces-400-mw-orders-us-uk>, 2013.

[33] S. Suryanarayanan, E. Kyriakides, Microgrids: am emerging technology to enhance power system reliability. IEEE: the expertise to make smart grid a reality, 2012.

[34] N. Thirumurthy, L. Harrington, D. Martin, L. Thomas, J. Takpa, R. Gergan, et al., Opportunities and Challenges for Solar Minigrid Development in Rural India, Technical Report NREL/TP-7A40-55562. <http://www.nrel.gov/docs/fy12osti/55562.pdf>, 2012.

[35] B. Torre, Microgrid Developments, Obstacles and the "Solar Happy Hour" UCSD Center for Research. <http://scalinggreen.com/2014/08/bill-torre-of-the-center-for-energy-research-on-microgrid-developments-obstacles-and-the-solar-happy-hour/>, 2014.

[36] The Borrower's Guide to Financing Solar Energy Systems, A Federal Overview, second ed., DOE/GO-10099-742, This Publication Is a Revision of DOE/GO-10098-660, <http://www.nrel.gov/docs/fy99osti/26242.pdf>, 1999.

[37] USAID: REMMP Briefing Note In-House Asset Finance for Small-Scale Renewable Energy. <http://www.arcfinance.org/pdfs/pubs/REMMP_Briefing_Note_In-House_Asset_Financing.pdf>, 2014.

[38] K. Valentine, Making solar affordable to those who can't afford, Clean Technica. <http://cleantechnica.com/2013/08/27/making-solar-affordable-to-those-who-cant-afford/>, 2013.

[39] B. Vergetis, B.V. Lundin, CESA: combining solar with energy storage the future of clean energy, Fierce Energy. <http://www.fierceenergy.com/story/cesa-combining-solar-energy-storage-future-clean-energy/2014-08-1>, 2014.

[40] B. Washom, J. Dilliot, D. David Weil, J. Kleissl, N. Balac, W. William Torre, et al., Ivory Tower of Power, IEEE Power & Energy Magazine (2013). 1540-7977/13/$31.00©2013IEEE.

[41] S. Zachary, 40 Companies & Organizations Bringing Solar Power to the Developing World. Renewable Energy World.com. <http://www.renewableenergyworld.com/rea/blog/post/2014/11/40-companies-organizations-bringing-solar-power-to-the-developing-world>, 2014.

Chapter 7

Solar Thermal Electricity and Solar Insolation

7.1 SOLAR THERMAL ELECTRICITY

Electricity can also be generated by concentrating solar energy into heating a fluid and producing steam which is then used to power an electrical generator. Unlike the solar photovoltaic method, this is an indirect form of generating electricity which uses mirrors to concentrate sunlight on a fluid to produce steam and then use steam turbines to run electrical generators. This method of generating electricity is commonly referred to as concentrated solar power (CSP). The main difference between CSP and solar photovoltaic is that the CSP has built-in storage capabilities, which allows it to generate electricity on demand after sunset and cloudy days. CSP can also be much larger in scale than solar photovoltaic.

CSP systems range from small standalone systems of a few kilowatts up to grid-connected power plants of several hundred megawatts. These systems work best in the bright sunny locations and when they are large in size because of the cost advantage in operation and maintenance. Small CSP systems producing 3–25 kW of power can be used as a source of distributed standalone power systems in the aftermath of disasters and supplying electricity to remote areas.

It is expected that CSP systems will become highly competitive in the power industry in the near future because of the decrease in the cost of implementation due to economies of scale, improvement in manufacturing and performance, as well as new development aiming at reducing the costs of operation and implementation.

The use of sun's energy to heat liquids was first employed to run a steam-driven printing press which was demonstrated by a French engineer/mathematician and physicist, at the 1878 World Fair in Paris, France. Augustin Mouchot used a 20 m^2 parabolic concentrating reflector that boiled water and produced steam to run the steam driven printing press. Between 1870 and 1880, Augustin Mouchot and his assistant, Abel Pifre, produced a series of machines ranging from the solar printing press, solar cooking, and solar engines driving refrigerators. In USA's south west, solar energy was used to heat water for homes during

Standalone Photovoltaic (PV) Systems for Disaster Relief and Remote Areas.
DOI: http://dx.doi.org/10.1016/B978-0-12-803022-6.00007-1

the late 1800s. In Los Angles, California, a quarter of residents at one point relied upon the sun to heat their water with rooftop solar thermal systems.

In 1910, the first installation of thermal energy equipment was used in the Sahara Desert to run a steam-powered engine, which was run by steam produced by sunlight. The project was abandoned due to inexpensive liquid fuel but was revisited several decades later.

At the beginning of 20th century Frank Shuman of United States used the principle of using parabolic concentrating collector with a steam boiler mounted at the focus using large parabolic trough collectors. He then built the system using parabolic trough collectors measuring 80 m (length) × 4 m (width) with a finned cast iron pipe at the focus to carry away stream into an engine. In 1913, this system produced 55 horsepower.

The first large experimental solar thermal electricity generation using high temperatures known as CSP plants started in United States during the early 1980s. Most of these plants used a large array of heliostats on the ground which focus the sun's rays onto a central receiver at the top of a tower. This type of system, called "power towers" uses a chamber as a central receiver at the top of a tower where the steam can either be produced directly, or uses a heat transfer fluid such as molten salt or mineral oil. The heat transfer material is then pumped away after attaining a high temperature to generate steam at the ground level where it is used to drive a turbine to generate electricity. In 1981, a 10 MW plant, called Solar One was built at Barstow, California. This plant was rebuilt in 1995 as Solar Two, using molten salt at 500°C. The plant included heat storage enabling it to produce electricity on a 24-hour basis.

In Spain, two commercial plants, Planta Solar 10 (PS 10) and Plata Solar 20 (PS 20) were commissioned in 2007 and 2009 near Seville, Andalusia. Planta Solar 10 is the first completely commercial solar central-receiver grid-connected system producing 11 MW of electricity. PS 20 is a 20 MW plant built next to PS 10 in 2009 with significant improvement to its predecessor's receiver with a capacity of more than 40 GWh of energy each year. It is capable of supplying power to more than 1000 homes. These CSP systems are developed and installed by Abengoa Solar.

In United States, the Ivanpah Solar Electric Generating system in the California Mojave Desert in 2014 became the largest CSP plant in the world with a capacity of 392 MW and capable of supplying power to 94,000 average American homes. The project was developed by BrightSource Energy and Bechtel and most of the power will be sold under the long-term power purchase agreements to Pacific Gas & Electric and Southern Californian Edison Company.

During the 1980s and 1990s, the cost of building, operating, and maintaining solar thermal electric systems decreased around 10% and is expected to drop further. With this trend, it is predicted that the solar thermal electric capacity will increase and in some countries it may be economically competitive with conventional electricity generation technologies. In United States, it is estimated by the National Renewable Energy Laboratory (NREL), that solar thermal power could provide 10% of demand. In emerging economies, the built-in storage capabilities of CSP will be used to generate electricity when demand peaks after sunset.

7.2 ADVANTAGES OF CSP

- Clean, reliable, and uses domestic renewable energy
- Flexible and dispatchable on demand
- Easily integrated into power grid
- Environmentally sustainable
- Reduces carbon emission and improves air quality for local communities
- Creates jobs and boosts national economy
- Can be located at nonagricultural and inhospitable places like deserts and barren sites
- Uses established technology like mirrors, turbines, tubes, electrical generators, etc.
- Generates electricity during the day when air-conditioning loading is high
- Clean-running daytime power plants help supplement other primary electrical generation sources
- Closely matches demand when integrated with thermal storage
- Well suited for use in hybrid configuration with fossil fuel plants especially natural gas combined cycle plants
- Uses minimum water when coupled with dry cooling

7.2.1 Disadvantages

- Expensive to install in remote areas
- Takes longer time to pay for itself
- Less reliable during night and cloudy days
- Cannot be economically used in a backyard to power a house
- Needs continuous maintenance

7.3 PRINCIPLES OF CSP SYSTEMS

Solar thermal electricity in a CSP plant is generated in two stages. In the first stage, solar energy is captured in the collectors and is used to heat a working fluid which may be water or molten salt. The second stage deals with the energy transformation in which electricity is generated by allowing steam to run a turbine or an engine. It is generally realized by a conventional steam turbine based on the Rankine cycle which is based on a thermodynamic model to predict the performance of a heat engine that converts heat into mechanical work. In standalone CSP systems, a heat transfer fluid is heated as it circulates through the receiver in the collectors. The heated fluid runs through a heat exchanger to generate high-pressure steam which is fed to a separate circle to drive a conventional steam turbine. The consumed steam from the turbine is condensed into liquid ready to be heated again in the steam generator to complete the circle. CSP plants consist of the following components as shown in Fig. 7.1:

- Solar collectors
- Heat transfer fluid
- Thermal storage
- Turbine

7.3.1 Solar Collectors

A solar collector is a major component of the CSP system which absorbs the incoming sunlight and converts it into heat. This heat is then transferred to a fluid such as air, oil, or water which flows through the collector where it is carried from the circulating fluid to the thermal storage tank and can be drawn for use at night or cloudy days. There are two types of collectors for CSP systems.

The first type is the nonconcentrating collector in which the area that intercepts the sunlight is the same as the area that absorbs the solar radiation. These type of collectors are also referred to as a line-focusing system which tracks the sun position in one dimension and the whole solar panel absorbs global sunlight.

The second type is a concentrating collector that intercepts direct sunlight over a large area and focuses it into a small absorber area, thereby increasing the radiation flux. This system uses a concave reflective surface and tracking to intercept and focus all of the sun radiation. The concentrating system is also called point focusing system which realizes higher concentration ratio than nonconcentrating systems because their mirrors track the sun in two dimensions. Concentrator collectors have a bigger interceptor area than an absorber and are

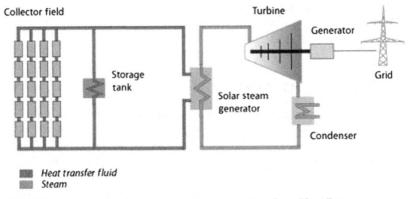

FIGURE 7.1 Components of concentrated solar power system. *Green Rhino Energy.*

more suitable for high temperature applications. Collectors can be classified into four different categories:

- Parabolic trough collectors
- Linear Fresnel reflectors
- Power tower reflector/receivers
- Parabolic dish reflectors

Among the four categories of collectors, the first two can be classified as nonconcentrating or line collectors and third and fourth can be classified as concentrating or point source collectors.

These collectors capture the sun's energy with mirrors that reflect and focus the sunlight into a receiver or absorber, and create the heat which is used to generate electricity. A heat transfer medium is used to generate steam at 400°C (750°F) which is used to drive a steam turbine generator. Most of these collectors comprise up to 40% of the total system cost for CSP technologies. In recent applications, the parabolic trough and power tower technology has become a maturing technology for providing a highly reliable operation and improved overall plant efficiency.

7.3.2   Heat Transfer Fluids

Heat transfer fluids carry the heat to the storage tank and then to the steam generator. As a result, it is important for good fluids to have a low viscosity and high thermal capacity. Water, synthetic oil, and molten salt can be used as a heat transfer fluids. Water is a good heat transfer fluid as it has a high thermal capacity and low viscosity. It is cheaper to use because its application in direct steam generation saves cost in the heat exchanger. However, it is unstable and difficult to manage at high temperatures.

Oil has a higher liquid temperature than water and has been a preferred choice to get around the high pressure issue. The problem with heavy oil is that hydrocarbons breaks down quickly if heated to within 400°C, therefore there are limits on the temperature at which the CSP can operate.

Molten salt which is a mixture of sodium nitrate and potassium nitrate can handle much higher temperatures than oil or water at 550°C allowing steam turbines to operate at greater efficiency. One problem with this heat transfer fluid is the risk of freezing of the salt in the longer length of receiver.

Steam was originally used in US Solar One plant which went into operation in 2007, but Solar Two utilizes molten nitrate salt because of its energy storage capabilities. The molten salt which is liquid at approximately 1000°C is kept in an insulated storage until the time it will be needed for heating up the water in the steam generator. The efficiency of energy storage using this method is approximately 99% where 1% of stored energy is lost due to imperfect insulation.

7.3.3 Thermal Energy Storage (Storage Tank)

The operation of a CSP plant can be extended beyond periods of no solar radiation by the addition of thermal energy storage (TES). This will make the system not only more dispatchable but it will also make it more cost effective and improve its performance by reducing the mismatch between supply and demand. There are two kinds of TES commonly used for CSP plants.

The first TES system is called sensible heat storage, and is based on storing thermal energy by heating or cooling a liquid or solid storage medium such as water, molten salt, sand and rocks. When a substance used as the TES is heated, its internal energy increases and when it cools, that thermal energy is released to the neighboring locations. This is an indirect method of heating a large volume of material for storing a certain amount of energy required by the CSP plants. Sensible heat storage system although is relatively cheaper than latent heat storage, it requires proper design to discharge thermal energy at constant temperatures.

The second method of thermal storage for CSP plants is latent heat storage which uses phase change materials (PCMs) for a shift in phase when the material changes from solid to liquid and vice versa. The heat stored and retrieved during the phase change process of material is called latent heat and can be used to reduce the size and cost of the storage system. Latent heat storage is a direct storage system which has the advantage of high storage density and ability to store energy with only a small temperature variation. TES through PCM is capable of storing and releasing large amount of energy in a wide range of temperature applications. However, the higher thermal resistance provided by intrinsically low thermal conductivity by the use of PCM is presenting a major technology barrier for its implementation. Because of high temperatures and the pressure involved, it is costly and is used as a buffer storage for peak power.

Most of the current CSP plants use sensible energy storage in molten salt to store energy which requires large volume of salt for two-tank storage systems. Molten salt is a mixture of nitrate sodium nitrate ($NaNO_3$) and potassium nitrate (KNO_3) at a 60:40 ratio which melts when heated above 230°C. In the two-tank system, also called a direct system, the heat transfer fluid in the form of molten salt is placed in two separate tanks, one at a high temperature and other at a low temperature. Molten salt fluid from low temperature tank flows through the solar collector where it is heated by sunlight to a high temperature. The heated fluid then flows to the high temperature tank for storage. Fluid from the high temperature tank flows through a heat exchanger where it generates steam for producing electricity. The fluid exits the heat exchanger at a low temperature and returns to the low-temperature tank. This system, although simple and cost effective, involves significant physical and cost redundancy in its construction. The Solana CSP plant which started operation in 2013, at Phoenix, Arizona, uses molten salt storage that can deliver power for nearly 6 hours after sunset.

7.3.4 Turbine

A steam turbine is a form of steam engine that extracts thermal energy from pressurized steam and converts it to rotary motion which is used to drive an electrical generator. A solar turbine works on the same principle as any steam-driven generator powered by the fossil fuels except the way the steam is produced to power the turbine. In a solar turbine, steam is generated by using a transmission fluid that is heated by capturing sunlight with a number of parabolic mirrors which in turn boils the water. The most important difference between powering steam turbines by fossil fuels and solar energy is the operation cycle. Due to the intermittent nature of solar radiation, solar turbines need to work efficiently during repeated starts and stops throughout the day. As a result, steam turbines for CSP plants should match the applications specific demands including a number of starts, rapid-startup capabilities and re-heat options for maximum performance.

One weakness of the current solar turbine is that transmission fluids cannot be heated above 400°C, although turbines are capable of operating with steam heated up to 540°C which would generate more power. This shortcoming could be overcome by placing the turbine on a high tower with the mirrors aimed to focus the sunlight directly on the steam boiler instead of using transmission fluid in the pipes. The majority of CSP plants except the parabolic dish operate according to Rankine thermodynamic cycle, wherein a steam turbine coupled to an alternator converts thermal energy into electricity. Steam turbines are only practical for very large CSP installations.

7.4 CSP AROUND THE WORLD

Global installed capacity of CSP plants in the last 10 years has jumped from 3.55 GW in 2005 to nearly 5 GW by the end of 2015, making it an established technology in the developed and emerging markets. Additionally, CSP activity is rapidly increasing to approximately 14.5 GW in various stages of development in 20 countries of the world. According to the *CSP Today Markets Forecast Report*, the CSP global installation will temporarily drop in 2016 but the activity will pick up again in 2017 with 1.2 GW of new installed capacity set to come online. In 2018, there will be even stronger growth, partly due to large projects planned in Egypt, Kuwait, and Tunisia. The report predicts that growth in global CSP capacity will rise from the current 4.7 GW (in 2015) to reach a capacity of between 10 and 22 GW by 2025 considering a pessimistic, conservative, and optimistic scenarios in that order.

New markets in CSP technology are emerging in most continents and locations where there is plenty of sun and skies are clear such as North Africa, South Africa, Americas, Australia, China, India, and Middle East. Spain, as of Jan. 2014, had a total CSP capacity of 2.3 GW making it the largest country in installing CSP plants. In 2013, Abu Dhabi installed Shams-I, which is one of the largest CSP project in the world. In United States, the Department of Energy and committed industry partners connected the most innovative five CSP plants to the grid in 2013. These five plants with a total capacity of 3.73 GW will quadruple the preexisting CSP capacity in the United States.

According to another report released by Transparency Market Research, the global CSP market is growing fast to reach US$8.67 billion by 2020 after being recorded around US$2.51 billion at the end of 2013. The report further estimates that the global CPS market is expected to progress at a compound annual growth rate (CAGR) of 19.4% between 2014 and 2020 in terms of value and 20.3% CAGR for the same forecast period in term of volume. The 2014 International Energy Agency (IEA) technology road map, foresees share of global solar thermal electricity to reach 11% by 2050. The road map also envisions that achieving 1000 GW of installed CSP capacity by 2050 would avoid the emissions of up to 2.1 Gt (Gigatonnes) of carbon dioxide (CO_2) annually.

NREL has been maintaining a database of CSP projects around the world in conjunction with SolarPACES (solar power and chemical energy systems) which is an international program of the IEA. The CSP projects are categorized into plants that are either operational, under construction, or under development. The profiles of the plants also include information on the type of technologies used for electricity generation, background information, a listing of project participants and data on the power-plant configuration. Solar thermal electricity generation systems consist of four types of CSP technologies including parabolic trough, power tower, linear Fresnel reflector, and dish/engine systems. The number of different types of CSP projects is given in Table 7.1.

Countries around the world have a total of 126 CSP projects. These projects include 90 operational projects, 9 under construction and 9 under development projects. Of the total CSP projects, 93 belong to parabolic trough, 21 are power tower, 12 are Fresnel reflector type and 2 are dish/engine projects. Table 7.2 lists the countries with the number of projects.

TABLE 7.1 CSP Projects Statistics

Number of countries with CSPs	20
Total CSP projects	126
Number of parabolic trough	90
Number of power tower	21
Number of Fresnel reflector	12
Number of dish/engine	2

TABLE 7.2 CSP Plants Worldwide

Name of the Country	Number of CSP Plants	Name of the Country	Number of CSP Plants
Algeria	1	Australia	4
Canada	1	Chile	2
China	4	Egypt	1
France	3	Germany	1
India	9	Iran	1
Israel	2	Italy	3
Mexico	1	Morocco	7
South Africa	7	Spain	50
Thailand	1	Turkey	1
United Arab Emirates	1	United States	26

7.4.1   Largest CSP Plants of the World in Operation

According to Zachary Shahan of Clean Technica, the largest solar thermal power plants in operation as of 2014 are:

1. Ivanpah Solar Electric Generating System in the Mojave Desert of California, United States—392 MW
2. Solar Energy Generating Systems in California, United States—354 MW
3. Solana Generating Station in Arizona, United States—280 MW
4. Solaben Solar Power Station in Logrosán, Spain—200 MW
5. Solnova Solar Power Station in Seville, Spain—150 MW
6. Andasol Solar Power Station in Granada, Spain—150 MW
7. Extresol Solar Power Station in Torre de Miguel Sesmero, Spain—150 MW
8. Shams-1 in Abu Dhabi, UAE—100 MW (largest single-unit solar power plant in the world)
9. Palma del Río Solar Power Station in Córdoba, Spain—100 MW
10. Manchasol Solar Power Station in Ciudad Real, Spain—100 MW
11. Valle Solar Power Station in San José del Valle, Spain—100 MW
12. Helioenergy Solar Power Station in Écija, Spain—100 MW
13. Aste Solar Power Station in Alcázar de San Juan, Spain—100 MW
14. Solacor Solar Power Station in El Carpio, Spain—100 MW
15. Helios Solar Power Station in Puerto Lápice, Spain—100 MW
16. Termosol Solar Power Station in Navalvillar de Pela, Spain—100 MW

7.5 SOLAR THERMAL TECHNOLOGIES

There are four different types of solar thermal technologies (also known as CSP) for generating electricity:

- Parabolic trough systems
- Power tower systems
- Linear Fresnel reflector systems
- Dish/engine systems

7.5.1 Parabolic Trough Systems

Parabolic trough systems as shown in Fig. 7.2 are the most cost effective and widely used systems for generating electricity in the world. It accounts for about 90% of the installed CSP base.

This system consists of a long parabolic-shaped collector with curved mirrors that focuses the sun's rays on a receiver pipe (absorber tube) located at the focal point of parabolic troughs. These troughs can be more than 600 m in length and the metal absorber tube is usually embedded into an evacuated glass tube to reduce heat losses. The troughs are rotated throughout the day as the sun moves from east to west to maximize the received solar energy. The metal absorber tube is filled with fluid, typically synthetic oil, which can be heated up to 400°C. Because of the parabolic shape, the troughs can focus the sun at 30–100 times its normal intensity. The fluid is then pumped through a heat exchanger that transfers heat to water which produces steam on boiling the water. The steam is used to run a turbine which generates electricity. The use of other heat transfer fluids like molten salts or direct steam allows the operation up to 550°C, hence improving the plant efficiency. These systems can also be designed as hybrids by using fossil fuel to back the solar thermal output during night time or during periods of cloudy days.

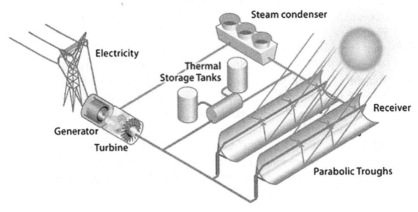

FIGURE 7.2 Parabolic trough system.

Most of the current thermal power plants for large power plants were developed by Lunz International which uses parabolic trough design called LUZ system (LS-1,2,3) collectors made from galvanized steel to support its torque based structure. Luz constructed nine Solar Electricity Generating Systems (SEGS) rated at 13–80 MW between 1984 and 1990. Two additional parabolic trough plants have been operational since the 1990s in United States at Nevada and Florida. Nevada Solar One (NSO) started in 2007 and generates 64 MW. Martin Next Generation Solar Energy Center, Indiantown, Florida, started its operation in 2010 with a capacity of 75 MW. This is the first hybrid combined cycle natural gas and CSP plant.

In Spain, more than a 1 GW of parabolic trough solar plants are built between 2007 and 2013 with the help of government subsidies in the form of feed-in tariffs. Most of these plants use molten salt for TES. The parabolic trough plants in Spain include 200 MW Solaben Solar Power Station, the 200 MW Solnova Solar Power Station and the 50 MW Andasol-1 solar power station which uses Eurotrough collectors.

The Solnova solar power station consists of five separate units of 50 MW each located at Seville in Solucar la Mayor which are built and operated by Abengoa Solar Company. The three units Solnova-1, Solnova-III and Solnova-IV were completed in 2010 in stages that started in 2007. Each unit uses 360 parabolic trough collectors based on ASTRO parabolic trough technology developed by the Solnova Solar Company. This technology involves long rows of curved heliostat mirrors that can be rotated toward the direction of the sun. The mirrors in turn reflect the sunlight which focusses onto a pipe containing a fluid. The three completed units are also equipped to support natural gas as its secondary fuel source. Fig. 7.3 shows Solnova-1 unit on the right, Solnova-III on the left

FIGURE 7.3   Solnova CSP plant using parabolic trough reflectors.

front and Solnova-IV on the rear left. The two towers in the background are PS10 and PS20 (see chapter: PV Systems Affordability, Community Solar, and Solar Microgrids and further details below).

In 2014, solar thermal power systems using parabolic trough technology include 354 MW SEGS plants in California, USA, the 280 MW Solana Generating Station at Arizona and 250 MW Genesis Solar Energy project at California.

The Solana solar generation system is developed and owned by Abengoa Solar which is a global company that currently has 1603 MW in commercial operation, 360 MW under construction and more than 320 MW in pre-construction in different countries. The plant as shown in Fig. 7.4 produces 280 MW (gross) solar output from two 140 MW steam turbine generators; it is the largest parabolic trough plant in the world.

This system uses 3200 mirrored parabolic trough collectors and 2.2 million square meters of reflective area covering 3 square miles of solar field. Electricity is generated with conventional steam turbines. The plant uses molten salt for TES and can supply 6 hours of dispatchable energy after the sunset or cloudy days. Solana generation plant can provide clean electricity to power 71,000 homes while avoiding 5000 tons of CO_2 every year. The fossil fuel backup used in this plant is natural gas.

FIGURE 7.4 Solana generating station using parabolic trough reflectors. *Courtesy of Abengoa.*

7.5.2   Power Tower Systems

Power towers system as shown in Fig. 7.5 is a newer technology compared to the parabolic solar trough and is more suited for large-scale grid connected power plants to be economical. A power tower system consists of many large sun tracking flat mirrors that focus sunlight on a receiver at the top of a tower. These tracking mirrors, numbering in hundreds to thousands, called heliostats can concentrate as much as 1500 times that of energy coming from the sun. This enormous amount of energy produces high temperature of 500–1500°C which can then be used to heat a fluid such as water or molten salt. The heated fluid in the receiver is used to generate steam which in turn is used to move the turbine generator for producing electricity.

Solar tower technology is among one of the solar thermal technologies that has the ability to store energy which enables it to supply electricity to the grid when needed and during cloudy weather or nights when there is no sun. It is a good source of providing large-scale distributed power to the nonproductive land like a desert without producing any greenhouse gases. It is estimated that a single 100-MW power tower with 12 hours of storage requires only 1000 acres of land for supplying electricity to 50,000 homes.

In the United States the Solar One power tower plant near Barstow, California, produced more than 38 million kWh of electricity between 1982 and 1988. This project produced 10 MW of electricity using 1818 mirrors, each measuring 40 m² (430 feet²) in size covering total area of 72,650 m² (782,000 feet²). The plant was modified in 1995 to include thermal storage capabilities using molten salt and was converted into Solar Two, by adding a second ring of 108 large heliostats measuring 95 m² (1000 feet²) of area. The current Solar Two contains nearly 2000 heliostats sun tracking mirrors with a total area of 82,750 m²

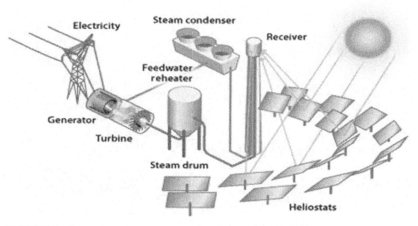

FIGURE 7.5   Power tower system.

(891,000 feet²) which reflect sunlight onto a receiver located at the top of a 300 foot tower. This gives Solar Two the ability to produce 10 MW power which can power an estimated 7500 homes. A digital control system is used to focus the heliostat to the receiver.

Spain has several power tower systems such as Planta Solar 10 (PS 10) and Planta Solar 20 (PS 20) which are water/steam systems with capacities of 11 and 20 MW, respectively. PS 10 is the first completely commercial solar central-receiver grid connected system producing 11 MW of electricity that started operation in 2007. It uses 624 glass–metal heliostats, pressurized water thermal storage system, saturated steam receiver and a turbine. The plant's thermal storage system has a 50-minute capacity at 50% load to handle cloud transients.

Solar's Planta 20 (PS 20) as shown in Fig. 7.6 is a 20 MW plant built next to PS 10 in 2009 with significant improvement to PS 10 receiver and was the largest solar concentrated power plant before the construction of Ivanpah Solar Electric Generating system in the United States.

PS 20 consists of solar field of 1255 mirrored heliostat with a surface area of 120 m² (1300 feet²) which reflects sunlight onto the receiver located on the top of 165 m (541 feet) high tower. Heat is produced when rays of sun reach the receiver producing high temperature and high pressure steam which is used to run turbines for generating electricity. PS 20 has the capacity of generating more than 40 GWh of energy each year. The plant has the capacity of supplying enough power to 10,000 homes by generating more than 40 GWh of energy each year.

The Ivanpah Solar Electric Generating System as shown in Fig. 7.7 is the largest solar power plant which has the capacity of producing 392 MW (gross) of electricity.

The project was developed by BrightSource Energy, constructed by Bechtel in partnership with NRG Energy, Google and the US Department of Energy. It consists of three separate units and is located in Ivanpah Dry Lake, California.

FIGURE 7.6 Solar's Planta (PS 20) solar plant using power towers.

FIGURE 7.7   IVANPAH power plant using a tower system. *BrightSource Energy.*

Ivanpah-1 has a total capacity of 126 MW and Ivanpah-2 and Ivanpah-3 are both 133 MW each and it is built on approximately 3500 acres. The plant deploys computer-controlled 173,500 heliostat pylons that track the sun in two dimensions and reflect the sunlight to the boilers which are located at the top of 450-foot-tall towers. Each heliostat consists of two mirrors and their aperture area is 15 m². The heliostat pylons are inserted directly in the ground that allows vegetation to coexist within the solar field below the mirrors. The plant uses dry cooling for steam condensation and 100% of the water used for steam powered turbines recycles back into the system.

The capacity of all three units can provide power to 140,000 homes in California during the peak hours of the day. The plant is zero-carbon generation source and will avoid 13.5 million tons of carbon emission over the first 30 years of operation.

The first commercial power tower plant which can dispatch power up to 10 hours after the sunset is recently installed at Tonopah, Nevada, United States. This plant is named the "Crescent Dunes Solar Energy Project" and was developed by SolarReserve and owned by SolarReserve's Tonopak Solar Energy LLC. This plant will have an output of 110 MW and uses molten salt for energy storage. The plant uses more than 10,000 heliostats which concentrate the sunlight atop a 640 foot tower that contains an external cylindrical receiver or heat exchanger. The heliostat aperture area is 115.7 m² and plant covers a 1500 acre field. It produces more than 500,0000 MWh of electricity annually and can provide power to 75,0000 homes during peak demand at night or cloudy days.

7.5.3 Linear Fresnel Reflector System

The linear Fresnel reflector technology received its name from the Fresnel lens which has multiple refracting planes designed to improve the concentration of light coming from many different angles onto a single point or line. This lens was developed by Augustin-Jean Fresnel in the 18th century, and allows a substantial reduction in the thickness, volume and weight of the lens but also reduces the quality of imaging. Giovanni Francia of Italy first applied it in 1960 for development of linear and a two-axis tracking Fresnel steam-generating system.

A linear Fresnel reflector's design is based on a principle between the power tower and parabolic trough concentrator systems. It is similar to parabolic trough system but contains fixed receiver pipe while mirrors track. The trough shape is split into multiple mirror facets. This mirror based system uses the same principle as a Fresnel lens uses for flat plane mirrors that track the sun to reflect light onto a tube. An additional secondary mirror is used in some systems behind the focal plane which directs the sunlight into the absorber pipe. The linear Fresnel reflector system shown in Fig. 7.8 uses long rows of flat or slightly curved mirrors to reflect sunlight onto a downward facing raised linear collector containing two stainless steel absorber tubes.

The collector in this system is a fixed absorber tube located at the common focal line of the mirror reflectors equipped with single or dual axis tracker to maximize the amount of sun energy collected throughout the day. A secondary concentrator is used to reflect the rays within the accepting angle. The Fresnel reflectors concentrate beam radiation to a stationary receiver. The receiver consists of two stainless steel absorber tubes. Each receiver has a secondary CPC

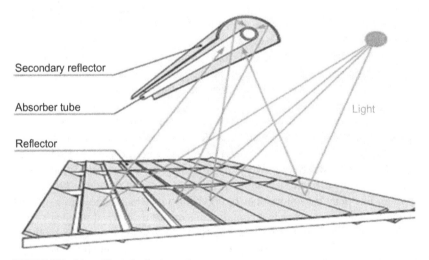

FIGURE 7.8 Linear Fresnel reflector system.

reflector that directs beam radiation on to the absorber tube. The entire optical system is enclosed in a sealed glazed casing. The absorber tube contains transfer fluid which is heated to create superheated steam that runs a turbine to produce electricity. New systems are designed that heat water to produce steam at 285°C (545°F) in the absorber tubes which will be directly used to drive a turbine to generate electricity instead of using a heat exchanger to produce steam from other high temperature fluids. The entire optical system is enclosed in a sealed glazed casing.

The system has a limited temperature-rising performance and is less efficient than other CSP systems but its simple design of ground-mounted mirrors makes it cheaper than the solar trough system. This system is ideal for small standalone systems in remote areas. Because of a fixed absorber and flat tracking mirrors, the system is structurally simpler and less expensive than other CSP systems. Linear Fresnel systems occupy less space than linear trough and power tower systems and they can be combined with fossil fuel backup for power generation during night or cloudy days. Compared with the parabolic trough technology, the energy yield per unit of area is much smaller. Yet the simple structure of the Fresnel collector and the possibility of using the space beneath the collector means that there is an important cost-lowering potential that can be exploited.

There are linear Fresnel reflector plants in operation or under construction in the world that produces more than 200 MW of power. The first commercial linear Fresnel reflector (LFR) system called Puerto-Errado 2 (PE 2) was completed in Spain in 2012 and built by Novatec Solar. PE 2 is a 30 MW plant with a solar boiler of 302,000 m^2 mirror surface area. The system is designed to produce approximately 50 million kWh of electricity every year being the world's largest CSP plant based on LFR. PE 2 consists of 28 rows of linear Fresnel collectors and solar field length of 940 m. The plant can provide clean energy to power 12,000 Spanish homes and avoid the equivalent of over 16,000 metric tonnes of CO$_2$ emission every year.

One disadvantage of linear Fresnel reflectors having only one receiver is that it is difficult to change the receiver's direction for a large size of the field required in a multimegawatt power generating system. This problem is solved in compact linear Fresnel reflector (CLFR) system which uses two parallel receivers for each row of reflectors. If the receivers in the CLFR are close enough, then individual reflectors can direct reflected solar radiation to at least two receivers allowing them to be more densely packed arrays and have lower absorber height. The receiver tubes in CLFR systems are placed at the focal point of a series of mirrors. An additional set of mirrors are also placed across the top of the tubes in order to capture and refocus any sunlight that does not directly hit the tubes. This configuration assures that all available light is used. Depending on the position of the sun, the reflectors can be alternated at different receivers to improve optical efficiency. Compared to linear Fresnel reflector system, CLFR systems can generate a large amount of heat and is cost effective in a

FIGURE 7.9   AREVA/Sandia Compact Linear Fresnel reflectors. *Randy Montoya.*

small land area. CFLR systems are scalable to the power plant needs and can be applied as standalone plants, as well as a booster in industrial applications.

AREVA solar has collaborated with Sandia National Laboratories to install a thermal energy storage system. This system is based on AREVA's CLFR and Sandia Lab's molten salt storage system, which uses molten salt instead of water as a working fluid in an elevated vacuum tube receiver. The thermal storage system draws molten salt from a 290°C tank and is heated to 550°C by the mirror's heat. The heated molten salt is passed to a separate tank for storage and is pumped through a heat exchanger that can be used later to produce electricity. The diagram of the system is shown in Fig. 7.9.

7.5.4   Parabolic Dish Engine

Unlike other CSP systems, parabolic dish systems are smaller in size, more efficient, and uses gas as the working fluid to generate electricity by running a heat engine. These systems can produce electricity between 3 and 25 kW at 30% conversion efficiency, which is higher than other CSP technologies. Due to its size and ruggedness it can be used as a standalone system for supplying power to remote areas and in the aftermath of disasters. The system as shown in Fig. 7.10 consists of a concentrator and a power conversion unit.

The solar concentrator or dish gathers the solar energy coming directly from the sun and reflects most of it to the thermal receiver as solar heat. The power conversion unit contains the thermal receiver and the engine/generator whose function is to convert heat into mechanical energy which runs an engine to produce electricity. The thermal receiver is the interface between dish and engine/generator and absorbs the concentrated beams from the sunlight. This system uses a parabolic dish of mirrors that focuses the sunlight from a large area to a single point collector in front of the dish. The parabolic dish looks like a typical

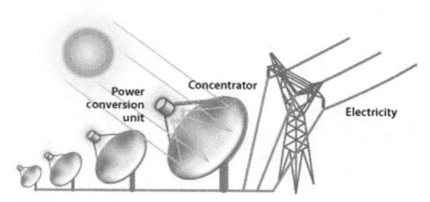

FIGURE 7.10   Parabolic dish/engine system.

satellite dish with a diameter of about 10 m for a 25 kW dish system at a the direct normal (maximum) solar insolation of 1000 W/m^2. The focused sunlight raises the temperature of the engine's heating chamber which is mounted at the focal point of the dish unlike the arrangement in a solar trough concentrator.

Most of the parabolic dish concentrator heating chambers contain a Stirling engine which uses hydrogen as a working fluid that is heated up to 1200°F in the receiver to drive the engine. This is achieved by converting heat to mechanical energy which compresses the cold working, heating the fluid, and then expanding it through a turbine (or moving pistons) to create mechanical power. Electricity is produced by coupling the engine to an electrical generator or alternator which converts mechanical energy into electrical power. The Stirling engine is air cooled and the whole system requires very little water to run and generate electricity. For higher efficiency the dish/engine uses concentrating collectors that track the sun to maintain optimum alignment with the sun. This is obtained by using two-axis tracking which focuses the sunlight on to a single point on a receiver so that the sun concentrates the solar energy at the focal point of the dish.

The parabolic dish system has a self-contained architecture and operates for longer periods with minimal maintenance. One disadvantage of the system is that it does not have any storage capabilities and the generated electricity has to be consumed immediately or transmitted to the grid. Cloudy days with no sun can result in lower and/or intermittent generation of electricity.

The world's first solar dish/engine was installed at Maricopa County near Phoenix, Arizona, in 2010 and was developed and owned by Tessera Solar as a demonstration project. The Maricopa solar project produced a capacity of 1.5 MW using 66 dishes and a dual-axis tracking dish Stirling engine. The project was decommissioned in 2011, and was purchased by United Sun Systems and partly by China.

FIGURE 7.11 Infinia solar thermal dish/engine.

Since the decommissioning of Maricopa project, Infinia Corporation has begun commissioning the first of seven commercial-scale PowerDishes in 2012 at the US Tooele Army Depot, Utah. The project is designed to produce 1.5 MW power using 4430 PowerDish units; the largest CSP project using Dish-Stirling technology as shown in Fig. 7.11.

The project is developed by Infinia Corporation and owned by Tooele Army Depot. Each PowerDish produces 3.5 kW of power and has an aperture area of $35\,m^2$. The plant uses helium as its heat transfer fluid and closed-loop cooling system. The project is modular in design and can be assembled by using mass-produced parts that are available off-the-shelf from hardware and automotive stores. The system can be easily realized as a standalone system in remote areas and in the aftermath of disaster for power generation.

7.6 SOLAR INSOLATION/RADIATION

In PV system design and analysis, it is important to know the amount of sunlight at the location of interest at a given time. The amount of sunlight or solar radiation is the combination (product) of hours of sunlight at a certain location and strength of the sunlight. This varies depending on the time of the year and the location due to the rotation of the earth. Solar radiation can be characterized by solar irradiance which is related to power per unit area and solar insolation which is related to radiant energy per unit area. Solar insolation is determined by summing irradiance over time and is expressed in kilowatt-hour per square meter spread over the period of a day (kWh/m^2 per day). It is usually used to rate the solar energy potential of a solar panel by multiplying it to the wattage of the panel which will be expected amount of daily energy for the solar panels.

The average daily solar insolation, also referred to as peak sun-hours, is roughly the amount of solar energy striking a 1 m² area perpendicular to the sun's location over a 1-hour period. Solar insolation as shown in Fig. 7.12 is the equivalent number of hours per day when the solar insolation averages 1 kW/m² so that the amount of solar energy is standardized at 1 kW hitting at 1 m² surface area.

It is useful to calculate peak sun-hours because electrical power output of PV modules are often rated at peak sun conditions of 1 kW/m².

$$\text{Peak sun hours / day}\left(\frac{\text{hrs}}{\text{day}}\right) = \frac{\text{Average daily irradaition}\left(\frac{\text{kWh}}{\text{m}^2.\text{day}}\right)}{\text{Peak sun}\left(1\frac{\text{kW}}{\text{m}^2}\right)}$$

The sun-hours of equivalent full sun is less than the amount of sunlight a site receives all day for two reasons. The first reason is due to the reflection of sun at a high angle relative to the PV panel because reflection loss depends on the latitude of the site where it is located. The second reason is due to the amount of the Earth's atmosphere through which sunlight is passing. At noon, the distance from the sun to the PV panel is the shortest, hence the sunlight passes through the least amount of atmosphere with least amount of reflection. Thus the PV panel produces the maximum power at noon, compared to the morning or evening when the sun is lower in the sky and sunlight passes through more atmosphere and has a greater angle of reflection. The most productive hours of sunlight are from 9 am and 3 pm with a clear sky.

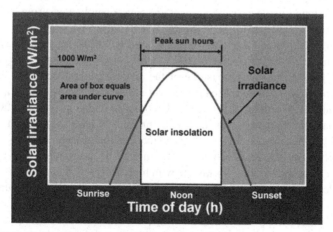

FIGURE 7.12   Solar insolation and peak sun hours. *NABCEP.*

Ground measurement of solar irradiation can be represented by global horizontal irradiance (GHI) which is an important parameter for photovoltaic designers and installers in order to calculate the electrical output of PV panels as well as for CSP. GHI is defined as the total amount of shortwave terrestrial irradiance received by a surface horizontal to the ground. GHI includes direct normal irradiance (DNI) and diffused horizontal irradiance (DIF).

$$GHI = DNI \times \cos(\theta) + DIF$$

where θ is solar zenith angle which is defined as the angle between the sun and a vertical line that goes straight up to the zenith.

DNI is the amount of solar radiation that comes in a straight line from the direction of the sun at its current position in the sky. It is considered to be received per unit area by the horizontal earth surface without any atmospheric losses such as scattering or absorption.

DIF is the amount of radiation that does not come in a straight line from the sun and without any shade or shadow. It is treated as received radiation per unit area by a surface which has been scattered by molecules and particles of air, aerosol and cloud in the atmosphere from all directions. On a sunny day without any clouds, the insolation is 100% GHI with 20% DFI and 80% DNI $\cos(\theta)$.

7.7 MEASUREMENT OF SOLAR RADIATION/INSOLATION

The total amount of radiation received by the photovoltaic module is the sum of direct irradiation, global, and/or diffused irradiation, and is measured in units of kW/m^2 as an instanteous power density. The measurements are taken periodically throughout the day as they vary depending on the time of the day, weather, and location of the site. Ground instrument measurement of solar radiation is performed by a device called radiometer which converts the sun's energy into electrical signals using a specific sensor and subsequently measuring the energy with conventional techniques. A pyranometer is used for measuring diffused or global radiation and a pyrheliometer is used for measuring direct radiation. A sunshine recorder, also known as Campbell–Stokes recorder, is also used to measure solar radiation. This is one of the earliest methods that was used for measurement; it is less accurate but also less expensive. For the sake of accuracy well-established locations have been collecting this data for more than 20 years.

Global solar radiation measured by pyranometers requires frequent calibration and maintenance because of the sensitivity of the instruments. It is difficult and costly to collect solar radiation data in case of large number of sparsely distributed ground stations worldwide with high accuracy. It also means that the spatial density of stations is far too low although the availability of ground solar radiation is on the rise.

An alternative approach for measuring solar radiation by existing networks of ground-based stations is through satellite-based images that have been developed over the last few decades. Most of the metrological geostationary satellites used for deriving solar radiations are equipped with imaging sensors that deliver spatially continuous information for large areas. These satellites orbiting at about 36,000 km are capable of offering a spatial resolution of up to 1 km and temporal resolution of 15 minutes. Determination of solar radiation reaching at the earth's surface is based on the idea that surface radiation is inversely related to the top of atmosphere reflectance which indicates the amount of clouds. The satellite images collected by the meteorological satellite over a large area with high time resolution helps to identify and forecast the cloud evolution which are then processed to the prediction of spatial variability of solar radiation at the ground level. It has been shown that compared to ground measurement, satellite-derived hourly irradiation is a more accurate option for locations that are situated more than 25 km from the ground station. Solar radiation/insolation for a particular location can be measured using:

- Pyreheliometer
- Pyranometer
- Campbell–Stokes recorder
- Satellite-based images

7.7.1   Pyreheliometer

A pyreheliometer is a broadband instrument that measures direct beam irradiance by pointing it permanently toward the sun. To keep the instrument aimed toward the sun, a two-axis tracking mechanism is often used. The sunlight enters the instrument window and is directed onto a thermopile detector which converts the heat into electrical signal. The pyreheliometer's field of view (aperture angle) is limited to 5 degrees to exclude all diffused radiation from entering the instrument. This is achieved by the shape of the collimating tube with precision aperture and detector design. The instrument is fitted with a quartz window that acts as a filter to pass radiation between 200 and 400,000 nm in wavelength. The quartz window also protects the instrument. The temperature of the detector is compensated to minimize sensitivity of fluctuation in ambient temperature. A pyreheliometer is a useful tool in assessing the efficiency of PV panels and in meteorological research.

Fig. 7.13 is an example of Eppley normal incidence pyrheliometer model sNIP. The Eppley normal incidence pyrheliometer model sNIP consists of a wire wound thermopile at the base of a tube with a viewing angle of approximately 5 degrees. This viewing angle limits the radiation that the thermopile receives to direct solar radiation making it normal incidence pyrheliometer. This instrument has a spectral range between 250 and 3000 nm with an output of 0–10 mV. The sensitivity of the model sNIP is rated at approximately 8 μV/W per m^2 with an impedance of approximately 200 Ohms.

FIGURE 7.13   Eppley Normal Incidence Pyrheliometer Model sNIP.

7.7.2   Pyranometer

A pyranometer is a broadband instrument that measures total global solar irradiance incoming from all directions including both direct and diffused in watts per square meter or kilowatt hours per square meter. It has a hemispherical (180-degree) view angle and its response to the incident sunlight is proportional to the cosine of the incident angle of the beam and a flat spectrum that covers wavelength of 300–5000 nm. The dome shape of pyrometer restricts the spectral response from 300 to 2800 nm from a field of view of 180 degrees. This hemispherical glass dome also shields the thermopile from wind, rain, and convection heat. Diffused radiation can be measured by blocking direct beam using arc-shaped metal discs which should be adjusted daily.

Two different types of sensors are used in pyranometers: the thermopile and photodiode or silicon solar cell. Thermopile-based pyranometers uses a strongly light-absorbing black paint that consumes all radiation from the sun equally which creates a potential that is proportional to solar irradiance. This produces the most accurate and most consistent response across a range of wavelengths.

Fig. 7.14 is an example of a Kipp & Zonen pyranometer Model CMP3 based on a thermopile sensor for measuring solar irradiance through the whole hemisphere with an 180-degrees field of view.

This type of pyranometer is a small, low-cost instrument which does not require any power to run. It produces a low voltage of 0–20 mV in response to incoming solar radiation and is designed for continuous indoor and outdoor use. The CMP3 pyranometer is intended for shortwave global solar radiation measurement in the spectral range from 300 to 2800 nm and measures irradiance up to 2000 W/m^2 with a typical sensitivity of 10 μV/W per m^2. Two CMP3s can form an albedometer that measures the reflectivity of the earth's surface.

A photodiode detector or silicon solar based pyranometer produces instantaneous electrical output from the incoming solar irradiance resulting in limited

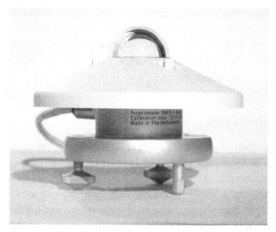

FIGURE 7.14   Kipp & Zonen pyranometer Model CMP3.

spectral sensitivity. These pyranometers are less accurate but are less expensive and more durable than the thermopile based pyronameter. Photodiode detectors or silicon based pyrometers are used in small handheld meters with a digital readout of solar irradiance.

Fig. 7.15 is an example of Kipp & Zonen Model SP Lite 2 pyranometer which uses photodiode detector to measure incoming radiation.

The Model SP Lite 2 pyranometer has a conical diffuser and 180-degree field of view which provides excellent directional (cosine) response. Its sensitivity is 60–100 µV/W per m^2 which is proportional to the cosine of the angle of incidence of the incoming radiation. This allows it to make consistent and accurate measurements. The overall spectral range of Model SP Lite 2 ranges from 400 to 1100 nm. The maximum solar irradiance of this pyranometer is 2000 W/m^2 and operational temperature range from −40°C to 80°C. Although the spectral

FIGURE 7.15   Kipp & Zonen Model SP Lite 2 pyranometer using photodiode detector.

range is limited by the photodetector, this model is particularly suited for monitoring the efficiency of a PV system where the spectral response is similar to PV cells. The shape of SP Lite 2 causes the rain to run off and can be used under all weather conditions. Two SP Lite 2 pyranometers can be easily bolted back to back to make a simple albedometer.

7.7.3 Campbell–Stokes Recorder

Another inexpensive method of measuring solar radiation is based on the Campbell–Stokes recorder, first invented in 1853 by Francis Campbell. It was refined in 1859 by George Gabriel Stokes. Although less accurate, it may be the best way of measuring sun hours in the aftermath of a disaster or in remote areas. This recorder consists of 4-inch-diameter annealed optical glass sphere that concentrates sunlight into a fine point on to a paper card inserted at the base of the unit. Each card is marked with a line produced as a result of heat from the sun indicating the hours of sunshine. Three types of cards are used to correspond to different seasons of the year. A transparent plastic template (optional) is available to help measure the curved trace more accurately. As the earth rotates, the presence of a resulting burn line indicates the period of sunshine. When the sun is obscured due to the clouds, breaks in the line (no burn) appears. At the end of the day the total length of the line, fewer breaks (no lines) is proportional to the duration of sunshine. The instrument is made strong enough to be placed outside throughout the year and is fully adjustable for operation at different geographical latitudes. In some sunshine recorders the number of hours in a day is measured when the sunshine is above a certain level such as $200\,mW/cm^2$. Data collected in this way can be used to determine the solar insolation by comparing the measured number of sunshine hours to those based on calculations and including several correction factors.

Fig. 7.16 is an example of Campbell–Stokes Pattern sunshine recorder Model 240-1070-L manufactured by NovaLynx Corporation.

FIGURE 7.16 NovLynx Cambell–Stokes pattern sunshine recorder. *Attribution: Bidgee, Bureau of Meteorology Campbell-Stokes recorder.*

The Campbell–Stokes Pattern sunshine recorder Model 240-1070-L uses a cut glass sphere to focus the sun's rays to an intense spot on curved cards which will char a mark. These cards are made from a special material which produces a clearly visible trace even in weak sunlight. The unit is housed in noncorrosive material in RAL 5009 (azure)/black color. It is integrated with circular spirit level on the base plate using additional mounting plate. Sunshine recorder 240-1070-L measures approximately $20 \times 18 \times 25$ cm ($7.84 \times 7.08 \times 9.84$ inches) and weighs 5.7 kg (12.566 lbs). The operating temperature for model 240-1070-L ranges from 25°C to 60°C in northern and southern equatorial zones.

7.7.4 Satellite-Based Images

The Heliosat-1, developed in the 1980s, is one of the earliest satellite-based methods for estimation of solar surface irradiance which is a measure of solar radiation at the surface of the earth. This method uses Meteosat (meteorological satellites) series of geostationary satellites which provide broad view of Europe, Africa and the Atlantic Ocean for meteorological purposes. It was widely used after several modifications. Heliosat-1 is currently operated by the European Organization for the Exploitation of Metrological satellites (EUMETSAT) and is initiated by the European Space Agency. EUMETSAT is an intergovernmental organization created through an international convention agreed by 30 European members with the primary objective of establishing, maintaining, and exploiting European systems of operational metrological satellites.

In 1992, the NREL and State University of New York (SUNY) in Albany released their first version of a National Solar Radiation Data Base (NSRDB) that contained hourly and half hourly values of global horizontal, direct normal, and diffuse irradiance as well as meteorological data. The 10 km gridded data set for 1998–2005 was derived from geostationary weather satellite images which covered whole of United States except Alaska above 60 degree N latitude. The model used for this data set uses hourly radiance images for daily snow cover data, and monthly averages of atmospheric water vapors, trace gases, and the amount of aerosols in the atmosphere to calculate the hourly total insolation falling on horizontal surface. This database has since updated collected data from a sufficient number of locations numbering 239 sites for 1961–90 and 1440 sites for 1991–2005.

In the years 1991, 2005, and 2012, NREL migrated to new technologies and updated the NSRDB three time by implementing new technologies to the original database. The update includes data sets from METSTAT (Meteorological–statistical) model developed at NREL and the second from SUNY at Albany model. NSRDB's 1991–2005 database contains data from 1454 ground sites that are divided into three categories: class 1 with 221 sites; class II with 637 sites and class III with 596 sites. Class I is highest quality data, more than 99% of which is modeled and less than 1% of which is taken from ground instrument measurements.

In 1997, NASA released its original surface meteorology and solar energy (SSE) data delivery website which provided access to various parameters needed in the solar and wind industry. The solar and meteorological data in this website was derived from 1993 NASA/World climate research program version 1.1 of International Satellite Cloud Climatology Project. SSE was modified after consultation with the renewable industry and SSE 2 was released in 1999. This was followed by new releases with parameters that were tailored to the need of renewable energy industry. The data is generated using Goddard Earth Observing System version 4 (GEOS) with a resolution of 1.25 degree longitude and 1 degree latitude for assimilation and simulation. NASA has a large archive of over 200 satellite-derived metrological solar radiation parameters and the data is available on a 1 degree latitude by 1 degree longitude grid covering 64,899 regions of the entire globe.

In 2003, NASA developed a POWER (prediction of worldwide energy resource) project through an SSE data set. A new web portal SSE 6.0 was released which was formulated from NASA satellite- and reanalysis-derived insolation and meteorological data for the 22-year period from Jul. 1983 to Jun. 2005. New features were added relative to the previous version of SSE release 5.1 by increasing the solar and meteorological data spanning from 10 years to 22 years, deriving the data from an improved algorithm, basing the temperature data on higher spatial resolution and increasing the number of renewable energy related parameters.

The POWER project was accomplished by the NASA Science Mission Directorate's satellite and reanalysis research program. Parameters based upon the solar and/or meteorology data were derived and validated based on recommendations from partners in the energy industry. NASA's geostationary satellites (shown in Fig. 7.17) used for deriving solar radiation to collect images over

FIGURE 7.17 NASA's near real time satellites for data stream.

a large area can offer a temporal resolution of up to 15 minutes and a spatial resolution of up to 1 km.

7.8   ONLINE DATABASES FOR SOLAR RADIATION

There are more than 20 solar radiation databases which can be accessed online at http://www.photovoltaic-software.com/solar-radiation-database.php.

These databases can be used to estimate solar radiation or solar insolation for locations in the United States and the world. Given below is further explanation of four such databases:

- NASA SSE 6.0
- Solar GIS
- NREL Renewable Resource Data Center
- RETScreen

7.8.1   NASA's SSE 6.0 for Global Locations

SSE 6.0 contains new parameters formulated for assessing and designing energy systems that was based on recommendations by the renewable energy industry. The primary aim of this project was to synthesize and analyze data that is useful to the renewable energy industry on a global scale. The project's goal includes providing data to the renewable energy industry in quantities and terms compatible with these industries and design engineering tools at locations where ground data is not readily available. Global, regional, and site-specific solar radiation and meteorological data can be used to evaluate potential renewable energy projects in any part of the world. Results of the data are provided for 1 degree latitude by 1 degree longitude grid cells over the globe and it is more accurate than previous releases of SSEs. This data has an easy-to-use format for plugging into software tools to design, build and market new energy systems for making use of renewable energy sources. These new systems are not only more efficient but will also be more affordable especially in developing countries. Private companies are taking advantage of the NASA dataset to develop new programs for advanced PV systems. For companies working in remote areas, it is an excellent tool to provide global coverage. RETScreen and SolarSizer are among the first to take advantage of SSE website in the design of PV systems at any location in the world.

7.8.2   SolarGIS Web Services

SolarGIS is a web-based tool that can be used to estimate solar insolation and solar radiation for many locations of the world. It is a new generation of web services consisting of various online tools and software aimed to increase efficiency and reduce uncertainty in planning and performance of PV systems. SolarGIS consists of a high-resolution database of solar radiation, air temperature and other auxiliary parameters for Africa, Asia and Australian Pacific, Europe, North & Central America, and South & Central America. SolrGIS data can be accessed by subscription-based tools including iMaps, pvPlanner and pvSpot or climData shop (for individual site-specific data purchases). However, solar maps and free trial and free demonstration of their products are available. SolarGIS web services are based on a model which runs 24 hours a day and process data covering a period from 1994 to the present. The data is derived from various geostationary satellites including Meteosat Prime and Indian Ocean Data Coverage (IODC), situated in 5 positions in the orbit covering the whole earth. Air temperature is derived from the atmospheric model of European Center for Medium range Weather Forecasting (ECMWF) and NCE models. The database consists of the following parameters which are derived for both the fixed-mounted or sun tracking photovoltaic systems:

1. Solar and PV data
   a. Direct nominal irradiance/irradiation (DNI)
   b. Global horizontal irradiance/irradiation (GHI)
   c. Diffuse horizontal irradiance/irradiation (DIF)
   d. Global tilted/in-phase irradiance/irradiation for fixed and sun-tracking surfaces (GTI)
   e. Optimum Angle (OPTA) for PV modules on fixed mounted construction
   f. PV electricity yield (PVOUT)
2. Meteorological data
   a. Air temperature at 2 m (TEMP)
   b. Relative humidity (RH)
   c. Wind speed (WS) and wind direction (WD) at 10 m
3. Geographical data
   a. Landscape
   b. Terrain
   c. Population

SolarGIS includes the following subscription-based online tools:

1. iMaps: high-resolution global interactive maps
2. ClimData: an interactive and automated access to solar radiation and air temperature
3. PV planner: a PV performance simulator with a new concept of simulation algorithms and data formats
4. PV spot: a tool for performance evaluation and monitoring PV systems

7.8.3 NREL Renewable Resource Data Center

The NREL, through its Renewable Resource Data Center (RReDC), provides free online assessments of solar, wind, and other energy resources in the United States. The RReDC's Solar Radiation Data Manual for Flat Plate and

Concentrating Collectors provides monthly average, minimum, and maximum values from data collected from 239 sites around the United States. The data accounts for variations in site and weather conditions at various times of the year and south-facing array orientation over the 30-year period of 1961 through 1990. The compressed files containing individual PDFs for the manual and site data tables can be downloaded in three compression formats: PC, UNIX, and Macintosh. Maps derived of the data in the tables are also available for viewing. Solar Radiation Data Manual for Flat-Plate and Concentrating Collectors (http://www.nrel.gov/docs/legosti/old/5607.pdf) and the updated National Solar Radiation Database (http://rredc.nrel.gov/solar/old_data/nsrdb/1991-2010/) which holds 1991–2010 solar and meteorological data of 1454 sites for United States and its territories is available on the Internet.

Each data sheet includes important information about the selected site, such as elevation, latitude and longitude, average climate condition, and barometric pressure. The latitude of the site is used to determine array orientation and barometric pressure may be used in computing the corrected air mass value at the site. The solar resource information is arranged in tables and columns.

7.8.4   RETScreen

The solar insolation can be obtained by RETScreen which was developed by collaboration of the CANMET Energy Technology Center of Natural resources, Canada and NASA, USA. It was developed to produce data output useful to users of the RETScreen International Clean Energy Project Analysis software. This software includes a number of databases to assist the user including a global database of climatic conditions, obtained from 6700 ground-based stations and NASA's satellite data.

The ground-based stations for climate monitoring are situated in cities around the world and NASA global satellites are used to monitor populated regions in the remote areas where surface measurements are not available. The number of ground-based stations in these cities is less than NASA global satellite/analysis data locations. These locations are represented by dots in Fig. 7.18. The full data set and maps covering the entire surface of the planet are available from NASA Surface meteorology and solar energy data set website: http://eosweb.larc.nasa.gov/sse/RETScreen.

The RETScreen software suite consisting of RETScreen 4 and RETScreen plus and is a decision-support tool developed by RETScreen International with the contributions of numerous experts from academia, industry and government. The software, available in more than 35 languages, can be used worldwide to evaluate the energy production, emission reduction, financial viability, cost saving and risks for various types of Renewable energy and energy efficient technologies. It is provided free of charge by the Canadian government, and is used by more than 425,000 people in 222 countries and territories. It is available in a case-study based college/university-level training course and is a part

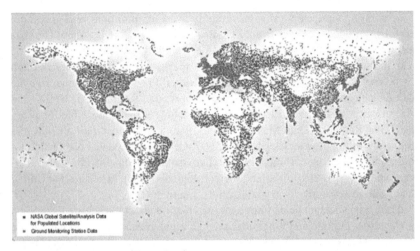

FIGURE 7.18   RETScreen Climate Database.

of the curriculum in more than 600 universities and colleges worldwide. More than 445,000 users in 222 countries and territories have used RETScreen as of Jul. 2015.

7.8.5   Estimating Solar Insolation for Global Locations Using the Photovoltaic Education Network

A quick way of estimating and comparing solar insolation for global locations can be obtained by accessing the following website:

http://www.pveducation.org/pvcdrom/properties-of-sunlight/average-solar-radiation#

Upon accessing this website, a world map will appear with about 50 marked locations as shown in Fig. 7.19.

These locations represent metropolitan areas in different part of the world including Asia, Africa, Australia, Europe, North and South America. In order to find solar insolation, one has to point the cursor to one of the locations on the world map. The result will be a bar graph showing the average daily radiation measured on the horizontal in units of kWh/m^2 per day for each month of the year. The graph in the figure shows solar insolation for Sydney, Australia. This is a useful online tool which can be used to get an estimate of solar insolation for different locations throughout the world. This online tool was developed by Christiana Honsberg and Stuart Bowden of Arizona State University, United States, and is part of an electronic textbook (available at http://www.pveducation.org) and on CDs. The material contained in the electronic book is based in part upon work supported by the National Science Foundation grants

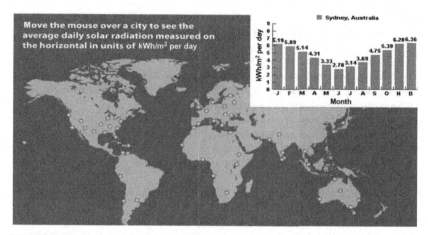

FIGURE 7.19   Average daily solar insolation/solar radiation for global locations. *Courtesy of pveducation.org.*

in collaboration with Quantum Energy and Sustainable Solar Technologies (QESST), University of Arizona and University of New South Wales, Australia. It is an excellent collection of resources for photovoltaic educators.

BIBLIOGRAPHY

[1] P. Blanc, B. Gschwind, M. Lefèvre, L. Ménard, L. Wald, Satellite-based estimation of surface solar irradiance. SPIE Newsroom (2011). <http://dx.doi.org/10.1117/2.1201105.003735>.

[2] M. Boxwell, Solar Electricity Handbook, Greenstream Publishing 2015. <http://www.solar-electricityhandbook.com/solar-irradiance.html>.

[3] W. Brooks, J. Dunlop, Photovoltaic Resource Guide (NABCEP). <http://www.nabcep.org/wp-content/uploads/2012/08/NABCEP-PV-Installer-Resource-Guide-August-2012-v.5.3.pdf>, 2012.

[4] G. Boyle, Renewable Energy: Power for a Sustainable Future, Oxford University Press, Oxford, 2012.

[5] Concentrating Solar Power Basics. <http://www.nrel.gov/learning/re_csp.html>.

[6] Concentrating Solar Power Dish/Engine System Basics. Energy. Gov. Office of Energy Efficiency & Renewable Energy. <http://energy.gov/eere/energybasics/articles/concentrating-solar-power-dishengine-system-basics>, 2013.

[7] Campbell–Stokes Pattern Sunshine Recorder Model 240-1070-L Manufactured by NovaLynx Corporation (nova@NovaLynx.com).

[8] CSP Today, Global CSP Capacity Forecast to Hit 22 GW by 2025. <http://social.csptoday.com/markets/global-csp-capacity-forecast-hit-22-gw-2025>, 2015.

[9] Concentrated Solar Thermal Systems. Green Rhino Energy. <http://www.greenrhinoenergy.com/solar/technologies/cst_systems.php>.

[10] A.F. Elmozughi, L. Solomon, A. Oztekin, S. Neti, P.C. Rossin, Encapsulated phase change material for high temperature thermal energy storage—heat transfer analysis, Int. J. Heat Mass Transf. 78 (2014) (2014) 1135–1144.

[11] Electricity Generating Systems: Large Scale Solar Thermal Plants Solar Thermal Power Plants Technology Fundamentals Published in Renewable Energy World (2003).

[12] Energy & Climate. <http://energy.sandia.gov/energy/renewable-energy/solar-energy/csp-2/csp-codes-and-tools/>.

[13] Electropaedia. <http://www.mpoweruk.com/solar_power.htm>.

[14] N. Ghumare, Transparency Market Research. <http://www.transparencymarketresearch.com/pressrelease/concentrated-solar-powermarket.htm>, 2015.

[15] R. George, E. Maxwell, High-resolution maps of solar collector performance using a climatological solar radiation model: Proceedings of the 1999 Annual Conference, American Solar Energy Society, Portland, ME, 1999.

[16] C. Honsberg, S. Bowden, A Collection of Resources for the Photovoltaic Educators. <http://www.pveducation.org/>.

[17] Infinia Inc.'s Modular Solar Thermal Dish. <https://www.mtholyoke.edu/~wang30y/csp/ParabolicDish_clip_image002_0001.jpg>.

[18] IVAPAH solar electric generating system <http://www.ivanpahsolar.com/>.

[19] IVANPAH project facts. Bright Source Limitless. <http://www.brightsourceenergy.com/stuff/contentmgr/files/0/8a69e55a233e0b7edfe14b9f77f5eb8d/folder/ivanpah_fact_sheet_3_26_14.pdf>.

[20] S. Kalogirou, Recent Patents in Solar Energy Collectors and Applications, Bentham Science Publishers Ltd., Higher Technical Institute, Nicosia, 2007.

[21] Kipp & Zonen Pyranometer Model CMP3. <http://www.kippzonen.com/Product/11/CMP3-Pyranometer#.Vjwko7erRgY>.

[22] Kipp & Zonen Model SP Lite 2. <http://www.kippzonen.com/Product/9/SP-Lite2-Pyranometer#.VjyBcrerRgY>.

[23] S. Kuravi, D.Y. Goswami, E.K. Stefanakos, M. Ram, C. Jotshi, S. Pendyala, et al. Clean Energy Research Center, University of South Florida, Tampa FL 33620.

[24] Linear Concentrator System Basics for Concentrating Solar Power. <http://energy.gov/eere/energybasics/articles/linear-concentrator-system-basics-concentrating-solar-power>.

[25] D.R. Mills, G. Robert, R.G. Morgan, Solar Thermal Electricity as the Primary Replacement for Coal and Oil in U.S. Generation and Transportation. <http://www.wired.com/images_blogs/wiredscience/files/MillsMorganUSGridSupplyCorrected.pdf>.

[26] NREL Concentrated Solar Energy Projects. <http://www.nrel.gov/csp/solarpaces/power_tower.cfm>.

[27] Power to the People: <http://earthobservatory.nasa.gov/Features/RenewableEnergy/renewable_energy.php>.

[28] Parabolic Dish System: Solar Energy Topics. <http://www.solarenergytopics.com/solar/parabolic-dish.html>.

[29] Z. Shahan, World's Largest Solar Power Plants. Clean Technica. <http://cleantechnica.com/2014/03/02/worlds-largest-solar-power-plants/>, 2014.

[30] Solar Power Towers Deliver Energy Solutions. <http://www.wipp.energy.gov/science/energy/powertower.htm>.

[31] Solar Insolation Map. <http://solarcraft.net/solar-insolation-map/>.

[32] SOLARRESERVE Solar Energy with Integrated Storage. <http://www.solarreserve.com/en/technology>.

[33] SOLARRESERVE Solar energy with Integrated Storage. <http://www.solarreserve.com/en/global-projects/csp/crescent-dunes>.

[34] Solar Turbines—Electricity Without Pollution, Advantage Environment. <http://advantage-environment.com/byggnader/solar-turbines%E2%80%94electricity-without-pollution/>.

[35] P.W. Stackhouse, R. Charles, W. Chandler, J.M. Hoell, D.J. Westberg, T. Zhang, et al. Using NASA's Near Real-Time Solar and Meteorological Data and RETScreen International plus Software for Monitoring Building Energy Systems at NASA Langley Research Center and Around the World 2013 American Metrological society meeting, Austin, TX, 2013.

[36] P.W. Stackhouse, D.J. Westberg, J.M. Hoell, Jr., W. Chandler, T. Zhang, "Surface Meteorology and Solar Energy (SSE)" Release 6.0 Methodology Version 3.1.2, Langley Research Center. <http://power.larc.nasa.gov/documents/SSE_Methodology.pdf>, 2015.

[37] P.W. Stackhouse, W. Chandler, D.J. Westberg, J.M. Hoell, C. Whitlock, T. Zhang, NASA applied science POWER Project, in 2011 RETScreen Annual Meeting, 2011.

[38] M. Šúri, T. Cebecauer, A. Skoczek, SolarGIS: solar data and online applications for PV planning and performance assessment, in 26th European Photovoltaics Solar Energy Conference, Hamburg, Germany, 2011.

[39] Surface Meteorology and Solar Energy (SSE) Release 6.0 Methodology Version 3.1.2. <https://eosweb.larc.nasa.gov/sse/documents/SSE6Methodology.pdf>, May 6, 2014.

[40] L. Stoddard, J. Abiecunas, and R. O'Connell, "Economic, Energy, and Environmental Benefits of Concentrating Solar Power in California." Black & Veatch of Overland Park, Kansas's subcontract report SR-550-39291for National Renewable Energy Laboratory. <http://www.nrel.gov/docs/fy06osti/39291.pdf>, April 2016.

[41] Solar Dish/Engine Systems. Sun Lab-snapshot. <http://www.nrel.gov/docs/fy99osti/23101.pdf>.

[42] Sun Lab/Snapshot Solar Dish/Engine Systems. <http://www.nrel.gov/docs/fy99osti/23101.pdf>.

[43] SolarPowerDish/EngineSystemBasicsOfficeofEnergyEfficiency&RenewableConcentrating. <http://energy.gov/eere/energybasics/articles/concentrating-solar-power-dishengine-system-basics>.

[44] Technology Road map: Solar Thermal Electricity, IEA (International Energy Agency). <http://www.iea.org/publications/freepublications/publication/technologyroadmapsolarthermalelectricity_2014edition.pdf>, 2014.

[45] Thermal Energy Storage for Concentrating Solar Plants. <http://research.fit.edu/nhc/documents/TES_NAI_Journal_Final.pdf>.

[46] US Department of Energy, The Year of Concentrating Solar Power. <http://energy.gov/sites/prod/files/2014/05/f15/2014_csp_report.pdf>, 2014.

[47] M. William, S. Wilcox, Solar Radiation Data Manual for Flat-plate and Concentrating Collectors. NREL/TP-463-5607, National Renewable Energy Laboratory, 1617 Cole, Boulevard, Golden, CO 80401, 1994.

Appendix

Appendix on PV Systems for Disaster Relief and in Remote Areas

APPENDIX A. BUILDING DYE-SENSITIZED SOLAR CELLS

Introduction

Dye-sensitized solar cells (DSSCs) are third generation thin-film solar cells which are one of the potential alternatives to conventional PV cells. DSSC was invented in 1991 by Brian O'Regan and Michael Gratzel, of Ecole Polytechnique Federale de Lausanne (EPFL) in France. Unlike a conventional c-Si cell, DSSC is based on a semiconductor formed between a photosensitized anode and an electrolyte and is called a photoelectrochemical cell. The advantages of DSSC over Si-based PV cells include high power conversion efficiency under cloudy and nondirect indoor light, low price-to-performance ratio, low cost of fabrication, transparency, low weight and mechanical robustness. These advantages mean that the use of DSSC to generate electricity has opened new markets in the remote areas of the world where more than billion people have no access to electricity and over 2 billion mobile phone users are without a reliable cell phone charging source. The solar-powered mobile phone chargers manufactured using commercialized dye-sensitized thin film technology will open new markets; the first already marketed in Africa. DSSC technology also offers quick and mobile power in the aftermath of disaster for providing light to homes, powering cell phones and other accessories. However, DSSC does have its weaknesses, such as relatively low efficiency, temperatures instability, high cost of components needed to fabricate, and low scalability.

DSSC consists of a porous layer of nanocrystalline titanium dioxide (TiO_2) covered with a molecular dye that absorbs sunlight as shown in Fig. A.1. The dye adsorbed on the surface of TiO_2 electrode acts as a sensitizer which absorbs the incident photon flux and is excited from the ground state to the excited state to allow for electronic conduction to take place. TiO_2 is a thick semiconductor nanoparticle film which acts as an electrode that provides a large surface area for the adsorption of light harvesting organic dye molecules. This porous electrode

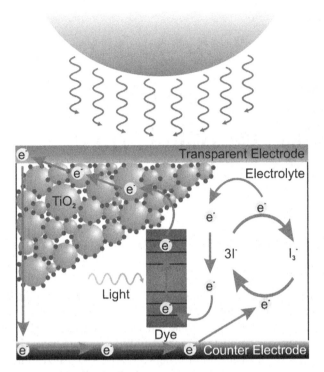

FIGURE A.1 Dye sensitized solar cell.

absorbs light in the visible region of electromagnetic spectrum and then initiates the charge separation process eventually leading to the generation of photocurrent. The process is accompanied by a charge transfer to the dye from an electron donor mediator supplied by an electrolyte, hence resetting the cycle.

DSSCs are easy to manufacture with traditional roll-printing techniques. These devices are semitransparent and semiflexible, allowing a range of uses that are not applicable to rigid photovoltaic systems. Most of the materials used to make these cells are low cost. However, a handful of more costly materials are necessary, such as ruthenium and platinum. There is a significant practical challenge involved in designing the liquid electrolyte for DSSCs, which must be able to remain in the liquid phase in all kinds of weather conditions. One of the problems with these cells are that only a few dyes can absorb a broad spectral range covering solar spectrum. The high efficiency of these cells is contributed to the fact that the mesoporous TiO_2 nanoparticles increase the surface area for dye chemisorptions to a thousand times over a flat electrode of the same size.

Building DSCC Solar Cells Using ICE Kit

To make DSSC on one's own is relatively simple but some of the components such as ruthenium dyes and platinum electrodes are expensive. So when making a DSSC cell as a DIY project, ruthenium dye can be replaced by natural dyes from fruits, and platinum cathodes can be replaced by carbon cathodes found in a pencil. A DIY kit to build DSSC cell (Gratzel cell) can be bought from Institute for Chemical Education for only US$45.00, plus shipping and handling. This prototype nanocrystalline dye sensitized solar cell kit was developed by Dr. Greg Smestad in 1998 using dyes from the plants. The kit is distributed (internationally) by the nonprofit organization, the Institute of Chemical Education (ICE) University of Wisconsin-Madison, Madison, United States. However all these components are locally available and a DSSC can be easily assembled following these instructions. The assembled DSSC solar cell lasts only a few months so it is meant to be used as a teaching tool and not a practical PV module. It is important that the procedure to assemble the kit should take place under the guidance of a faculty member who understands the basic chemical safety practices. The nanocrystalline solar cell kit shown in Fig. A.2 contains the following components which include all of the chemical supplies, nanoparticle titanium dioxide, and triiodide electrolyte to build five solar cells.

- Instruction manual
- Woodless HB graphite pencil
- 10 binder clips
- 10 conductive (tin oxide coated) transparent glass slides in glassine envelops
- 20 g colloidal titanium dioxide powder (TiO_2), Degussa P25
- 15 mL iodine electrolyte solution in a dropper bottle

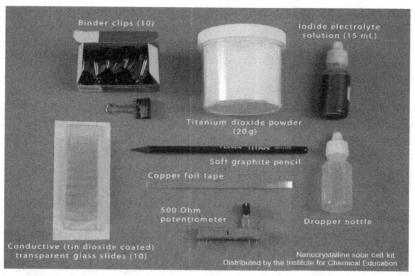

FIGURE A.2   Components of ICE's DSSC kit. *Copyright by Board of Regents of the University of Wisconsin System. Used with permission.*

- 500 Ohm potentiometer
- 12 cm copper tape
- 1 dropper bottle

To assemble the kit, fruits for dyes, some basic chemicals, and supplies are needed which are not provided in the kit. The accompanied manual first published in 1998 and revised in latest version of Oct. 2010 gives detailed information on building the cell, how they work and how to measure their electrical properties. Additional information can be accessed at the Sol ideas website: http://www.solideas.com/solrcell/cellkit.html. The following two sources are highly recommended.

- "Constructing a Dye Sensitized Solar cell." A YouTube video by Dr. Neal Abrams of Department of Chemistry, SUNY College of Environmental Science and Forestry.
    https://www.youtube.com/watch?v=17SsOKEN5dE, uploaded Oct. 2009.
- "Titanium Dioxide Raspberry Solar Cell." A YouTube video from University of Wisconsin-Madison Materials Research Science and Engineering (MRSEC)
    https://www.youtube.com/watch?v=Jw3qCLOXmi0, Published Aug. 2012.

Instructions for Building the ICE's Dye Sensitized Solar Cells

Building a DSSC is explained briefly in the following steps 1–4, Figs. A.3–A.6. While building a DSSC, safety should be strictly observed as you will be working with sharp glass, acids, heat sources, and staining dyes. Wear gloves during this lab work. If your skin comes in contact with any chemicals during the creation, wash the exposed areas with copious amounts of water. As always, goggles are required when working.

Step 1—Stain the Titanium Dioxide with the Natural Dye: Stain the white side of a glass plate which has been coated with titanium dioxide (TiO_2). This glass has been previously coated with a transparent conductive layer (SnO_2), as well as a porous TiO_2 film. Crush fresh (or frozen) blackberries, raspberries, pomegranate seeds, or red Hibiscus tea in a tablespoon of water. Soak the film for 5 minutes in this liquid to stain the film as shown in Fig. A.3. As a result, a deep red-purple color will appear in practice.

If both sides of the film are not uniformly stained, then put it back in the juice for 5 more minutes. Wash the film in ethanol and gently blot it dry with a tissue.

Step 2—Coat the Counter Electrode: The solar cell needs both a positive and a negative plate to function. The positive electrode is called the counter electrode and is created from a "conductive" SnO_2 coated glass plate. A Volt-Ohm meter can be used to check which side of the glass is conductive.

FIGURE A.3   Staining the TiO$_2$ film. *From G. Smestad, Education and solar conversion: Demonstrating electron transfer, Sol. Energy Mat. Sol. Cells 55 (1998) 157.*

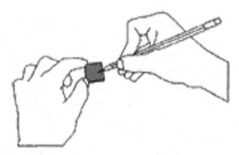

FIGURE A.4   Applying a thin graphite layer to the conductive side of plate's surface. *From G. Smestad, Education and solar conversion: Demonstrating electron transfer, Sol. Energy Mat. Sol. Cells 55 (1998) 157.*

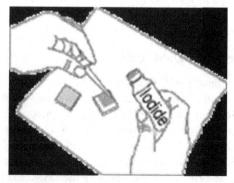

FIGURE A.5   Placing drops of the iodide/iodine electrolyte solution on the film. *From G. Smestad, M. Graetzel, Demonstrating electron transfer and nanotechnology: a natural dye-sensitized nanocrystalline energy converter, J. Chem. Educ. 75 (1998) 752.*

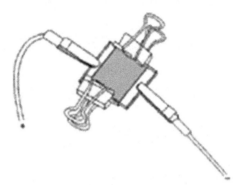

FIGURE A.6   Placing counter electrode on top of the stained film. *From G. Smestad, Education and solar conversion: demonstrating electron transfer, Sol. Energy Mater. Sol. Cells 55 (1998) 157.*

When scratched with a fingernail, it is the rough side. The "non-conductive" side is marked with a "+." Use a pencil lead to apply a thin graphite (catalytic carbon) layer to the conductive side of plate's surface as shown in Fig. A.4.

Steps 3 and 4—Add the Electrolyte and Assemble the Finished Solar Cell: The iodide solution serves as the electrolyte in the solar cell to complete the circuit and regenerate the dye. Place the stained plate on the table so that the film side is up and place one or two drops of the iodide/iodine electrolyte solution on the stained portion of the film as shown in Fig. A.5.

Then place the counter electrode on top of the stained film so that the conductive side of the counter electrode is on top of the film as shown in Fig. A.6.

Offset the glass plates so that the edges of each plate are exposed. These will serve as the contact points for the negative and positive electrodes so that you can extract electricity and test your cell.

Use the two clips to hold the two electrodes together at the corner of the plates.

Testing the Assembled DSSC Cell

Connect your assembled DSSC to a multimeter with alligator clips and check the voltage and current. Place the cell under an indoor light source, along a windowsill or in full sun that has some protection. Direct sun may cause some degradation. The output from your assembled DSSC is approximately 0.43 V and 1 mA/cm^2 when the cell is illuminated in full sun through the TiO$_2$. Make more than one DSSC and connect them in series to increase the voltage and in parallel to increase the current. Try connecting the DSSC to charge a rechargeable battery, make a solar charger, or power your calculator.

Perovskite Based Solar Cells

The early version of DSSC fabricated by O'Regan and Gratzel using a ruthenium-based dye yielded 7% power conversion efficiency which used 10-µm-thick porous TiO_2 nanoparticle films for the photoelectrode. This lower overall efficiency was obtained because of the inherent voltage loss during the regeneration of the sensitized dye. This efficiency did not change very much until the scientists at EPFL (Ecole Polytechnique Federale de Lausanne) developed a solid state version of the cell in 2013. By using a new two-step process the scientists at EPFL fabricated a DSSC which raised the efficiency to 15% without sacrificing stability. The new solid state embodiment of the DSSC used a perovskite material as a high harvester and the cell's electrolyte was replaced by an organic hole transport material. Fabrication of a typical DSSC using perovskite material involves deposition of entire material directly onto a metal oxide film. However adding the entire material onto a metal oxide in one step often causes wide variation in the morphology of the solar cell. This variation in morphology affects the efficiency of the solar cells which makes it difficult to use them in everyday applications. Michael Gratzel's team solved this problem by using a two-step process. In the first step, the team deposited one part of perovskite into the pores of metal-oxide scaffold and in the second step the team exposed the deposited part to a solution that contains the remaining components of the perovskite. When these parts come into contact, they react instantaneously and convert into the complete light sensitive pigment. This light-sensitive pigment permits much better control over the morphology of the solar cell. This method allowed DSSC to produce a high conversion efficiency with a high reproducibility.

Perovskite is a type of mineral that was first found in the Ural Mountains by Gustav Rose in 1839 and named after Lev Perovski who was the founder of Russian Geographical Society. Perovskite material, mostly composed of organic metal and halogens such as $CaTiO_2$, are easy to synthesize. Solar cells made from perovskite material steadily increased their efficiency from 3.8% in 2009 to more than 20% in 2015 for ultra-small device areas measuring $0.1\,cm^2$. Perovskites solar cell can be manufactured in a few steps without the use of costly equipment like the one used for silicon-based PV cells. Dense layers of crystallized perovskite are produced when a solution containing the electrode material coated onto a substrate evaporates, making it a very inexpensive way of making solar panels.

Solar cells made from perovskite material are the fastest advancing solar technology to date. With the potential of achieving even higher efficiencies than Si-based cells. It is predicted that in the future, solar power will not just come from bulky solar panels currently used on roof tops. According to a recent report, the solar panels of future will be lightweight, flexible, transparent and ultra-efficient. It will be possible to coat shingles, windows, or skylights with these new panels at a very low cost.

The new breakthrough in Perovskite cell technology has made it commercially attractive, with start-up companies promising to market the new modules by 2017. One such company is Solaronix (http://www.solaronix.com/documents/solaronix_materials.pdf) which was founded in 1993 after the discovery of a new generation of solar cells at the EPFL. Solaronix has been serving a worldwide customer base with end-to-end expertise in photovoltaics, from materials production to solar panels and test equipment manufacturing. The company also provide complete sets of components to help experimenters accelerate their work in the following areas:

- Perovskite solar cell kits
- Test cell kits (laboratory dye solar cells)
- Education cell kits (training and teaching dye solar cells)
- Demonstration cell kits (large solar cells)

APPENDIX B. ASSEMBLING ONE'S OWN PV MODULES

Introduction

Photovoltaic modules are the fundamental building blocks of PV systems and consist of PV cells which are connected in series and/or parallel to produce higher voltages, currents and power levels. The PV module also includes an encapsulant, a frame and a junction box. For protection, the solar cells are encapsulated in a piece of glass, EVA (ethylene vinyl acetate), a back sheet of TPT (tedlar polyester tedlar) and an aluminum frame around them. PV modules are manufactured by an automated process where each step is performed by using robust robots. Each of the robot-based tools works within a safety fence that excludes humans. However it is possible to make one's own PV modules from PV cells which are easily available from the manufacturers and other suppliers. These assembled PV modules, however will not match the reliability and durability of the commercially available modules from the manufacturers. Some of the reasons to make PV modules on one's own from scratch are:

- It is difficult and/or expensive to import large PV panels because of remote locations
- To provide electricity in an area without access to grid and start a cottage industry
- Require a hands-on approach to learn more about solar energy
- Require a special shape or power of PV modules not available from the manufacturer
- It is needed in a student lab or for research in an academic setting
- Has acquired inexpensive PV cells from a factory close out

Solar panels are made in three steps. The first step involves the process of tabbing and bussing which links individual solar cells together to form a PV module. These interconnected solar cells are then connected to a power output, which provides a method to transfer power from the solar cells to final output through a junction box. The second step is to encapsulate the interconnected cells by laminating onto glass. The encapsulants are designed to protect the PV module against weather and UV aging while ensuring the maximum amount of visible light transmission to PV cells. The third step is to frame the encapsulant and interconnected PV cells in order to protect against weather and other impacts. The frame should be light, inexpensive, and easy to install. Thin film solar cells made of copper indium diselenide (CIS), although less efficient, are much easier to assemble into solar panels. The method used to make PV modules in this section can be applied to both larger PV systems as well as to smaller systems like cell phone chargers with some modifications.

Interconnecting and Bussing of PV Cells

The standard PV cell measuring $3'' \times 6''$ produces output voltage of $0.5\,V$ and $3.5\,A$. In order to increase the voltage, PV cells are connected in series, and for increasing the current the cells are connected in parallel. A typical solar module uses 36 solar cells although panels with 30, 32, 33, or 44 cells are also available. A module of 36 cells with each cell equal to $0.5\,V$ connected in series produces $18\,V$. This arrangement is used to charge a $12\,V$ battery as the voltage is reduced when the cells get hot. A $12\,V$ battery typically needs about $14\,V$ for charging. A module of 36 cells has become the standard of the solar battery charger industry.

The first step of interconnecting PV cells and bussing is performed in two basic processes, the PV cell interconnecting by stringing and the PV module assembly by bussing. PV cell interconnection occurs when individual cells are electrically joined together using stringing or tabbing ribbon. The interconnected ribbon is soldered directly onto crystal silicon cell and carries the current generated in PV cells to the bus bar. The individual cells numbering 6–10 in a cluster are connected by stringing (tabbing) ribbon of typically $2\,mm$ wide which carries the PV cell's current. This current is then carried to a larger ribbon, the bus ribbon which carries power from the cell cluster to the junction box for final output.

In the PV cell interconnect, each cell is soldered with a tabbing wire by placing the cell on a flat surface. The length of tabbing wires for $3'' \times 6''$ PV cells should be 6 inches in length and a panel of 36 cells requires 72 of such wires. After the first step of soldering, half of the length of your tabbing wires sticks out of the first PV cells that you have already soldered. Hence the wires that stick out are connected to the back side of the next cell in a series connection. Care must be taken because PV cells are very thin and fragile. A tabbed PV cell is shown in Fig. B.1.

For stringing a PV cell, solder is applied to the 6 white squares on the back of the cell and is repeated for 4 strings of 9 cells. Stringing is defined as a group of PV units which are wired in series to increase voltage. It is advisable to use a board with 2 wooden strips acting as a guide. A cell is placed face down between the strips and its wires from the front side are then soldered to the 6 squares on the back

FIGURE B.1   A tabbed PV cell.

FIGURE B.2   3 PV cells face down.

of another cell. The front side of the cell is colored blue. This process is repeated for 9 PV cells, shuffling the strings up the board between the wooden strips (Figs. B.2–B.4) which is used to get a uniform spacing between cells. When the 9 cells are finished, two extra tab wires are attached to the back of the end (first) cell which has no wires on it.

PV bussing is achieved by joining together clusters of typically 20–80 PV cells in a module using 5 mm wide tabbing. PV cells' bus bar connects the interconnect ribbons to the junction box with a hot-dip tinned copper conductor installed around the perimeter of the solar panels. A soldering iron of 40–90 W with a rosin flux is used to connect PV cells together, by connecting the cell rows. The tabbing wire is first cut on the end-run solar cells to approximately 1/2″. The next step is to use some of the bus wire to connect the positive to negative at the end runs of the PV cells cluster. This will end up in 2 rows wired together at one end of the solar panel and 2 rows connected at the other end of the solar panel as shown in Figs. B.5 and B.6.

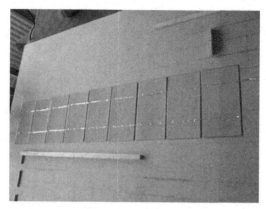

FIGURE B.3   9 PV cells face down.

FIGURE B.4   9 PV cells face up.

FIGURE B.5   Tabbing wire cut back to 1/2″.

FIGURE B.6 Connection of 2 rows of solar cells with bus wiring.

Junction Box

A solar junction box is installed at the back of every PV module and acts as an interface between the conductor ribbons on the module and the DC input and output cables. The main function of the junction box is to protect the solar panel from the hot spots and increase the life of the panel. The junction box as shown in Fig. B.7 consists of blocking and bypass diodes which are used to keep the electricity from entering the solar panel at the time of darkness or if the panels are covered by leaves or shade. It will also help keep the battery from draining unnecessarily.

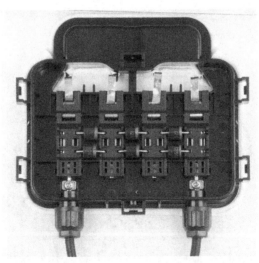

FIGURE B.7 Junction box. *By Elvis untot - Own work, CC BY-SA 3.0, https://commons.wikimedia .org/w/index.php?curid=19979692.*

Mounting and Sealing of Junction Box

Field testing experience shows that in a PV system, the components most prone to failure are the junction boxes connectors and cables on the rear of the PV panel. These components comprise 1–2% of the overall price of the cost of a PV system and can help to improve the efficiency of PV system and reduce the safety risks. These components should be mechanically robust, should have long expected lifetimes, and should be resistant to temperature extremes. The components should be approved for use in PV systems by an independent contractor. The junction box should be of good quality, rugged and should have a space for making terminal connections and bypass diodes in addition to the following requirements:

- Good adhesion to metal, glass and other components like backsheet laminate
- Reliable bonding at different conditions and range of temperature cycles
- Immune to deterioration from UV and extreme temperatures
- Negligible low moisture vapor transmission rate
- Easy inclusion in the manufacturing and assembly process

The following type of tapes are used to mount a junction box:

- Liquid adhesives
- Adhesive tapes.
- Acrylic foam tapes

Liquid Adhesives

Room temperature vulcanization (RTV) silicon is a type of liquid which has been used since the start of the PV industry. It can be applied both manually and as a part of automated dispensing process. Silicon RTV provides a strong durable bond and fulfils most of the above-mentioned requirements for a quality junction box. Since it can be incorporated into the automated inline manufacturing dispensing systems, it helps to reduce material cost and processing time with no error compared to manual application.

Despite many advantages, the RTV method of mounting and sealing has some disadvantages such as long curing time, inconsistency in application affecting durability, quality, and moisture vapor transmission rate. This process can be messy even with dispensing equipment.

Adhesive Tapes

In order to overcome some of the drawbacks of liquid adhesives mentioned above, some solar panels' manufacturers have suggested the use of adhesive tapes for mounting junction boxes. Many precut rolls of double-sided adhesive tapes including polyethylene foam tapes, polyurethane foam tapes and acrylic foam tapes have been used to save time required for the liquid adhesive to cure.

Polyethylene (PE) and polyurethane foam tapes made in different thickness and grades have been used for attaching junction boxes and sealing edge and frame by the PV manufacturers. These tapes with typical thicknesses of 0.8, 1, and 1.55 mm, with a thickness tolerance of ± 20%, are coated on both sides with adhesive using a transfer lamination process giving a variance in the performance of the tape. PE tapes are easy to use, cost effective, and can fit into solar panels' manufacturing processes. These tapes are very useful in gap filling and bond applications not subjected to a lot of stress.

However, PE tapes suffer from low compressive strength, can easily rupture with very little force and does not recover well from compression. The foam on the tapes will degrade and break down leading to leakage and water absorption when used in glass, aluminum, or plastic as a result of temperatures extremes. PE tapes have limited service life of about 9 years and is not appropriate for PV systems expected to last longer than 25 years.

Acrylic Foam Tapes

In order to overcome the drawbacks of PE tapes mentioned above, high strength acrylic tapes are used as an alternative for PV applications. An acrylic foam tape contains air bubbles and injected glass beads to give the tape a viscoelastic effect. Such a tape will stretch and retract to original shape without breaking the bond or adhesion loss, making it necessary capabilities of expansion/contraction for PV application. Acrylic foam tapes are capable of withstanding very high UV exposure and temperature extremes for long periods and can resist very high wind forces and snow loads. These tapes are used for junction box mounting as well as sealing of air, moisture, dust, and sealing for frame bonding, and edges. For a wide range of junction boxes and shapes, these tapes can be precision die-cut as a gasket. The biggest drawback of these tapes is the cost.

For mounting and sealing larger and heavier junction boxes, a combination of both liquid and tape adhesives are being used where the tape is recommended to hold the junction box in place while the silicone is applied to mount it.

Encapsulation

Encapsulation is the second step in the process of making the PV modules. This step is required to provide mechanical support for Si-crystalline PV cells which are fragile and can break under tension. Encapsulation should also isolate electrically conducting parts of the solar cells such as bus bars, terminals, and interconnects from exposure to the environment causing corrosion and deteriorating interaction. Encapsulation structure typically consists of tedlar at the bottom base, EVA encapsulant for the cells at the top and bottom, and low iron glass at the top front as shown in Fig. B.8.

Ethylene vinyl acetate (EVA) is one of the commonly used encapsulants and have been used since 1970. EVA is a fast-cure encapsulant that is designed to work with PV modules for protection against UV-aging and weathering while ensuring maximum amount of visible light transmission to solar cells. This copolymer film is an essential sealant of photovoltaic solar modules for ensuring the reliability and performance. TPT (Tedlar Polyester Tedlar) as a

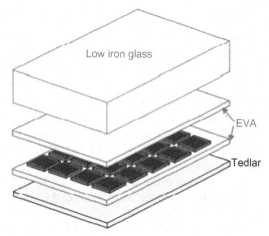

FIGURE B.8   EVA encapsulant for bulk silicon module materials. *Courtesy of pveducation.org.*

backsheet is a tri-layer structure normally made as a laminated film composite. This structure allows the fluoropolymer to protect both sides of the polyester from photodegradation. TPT film is tough, photostable, chemically resistant, and unaffected by long-term moisture exposure.

The use of organic EVA; however, requires addition of extra material to block UV radiation. This reduces the transparency at lower wavelengths which in turn reduces the overall efficiency of PV modules. Silicones on the other hand can be considered as a hybrid inorganic glass and organic linear polymer which has the potential of improving module efficiency, improved durability and faster processing times. Silicone polymers also have unique feature of low viscosity prior to curing which allows it to flow over and around the cell architecture. Inherent properties of silicones such as very low ionic impurities, low moisture absorption, low dielectric constant and broad temperature utility in conjunction with their excellent optical transparency over a wide spectrum and UV stability make them highly suitable for meeting the materials requirements for encapsulation of PV cells. They can be formulated to have low modulus and be stress relieving while also having excellent adhesion to the glass and PV cells and substrates. Dow Corning Sylgard 184 is made up of two parts; 10:1 silicone encapsulant is agitated gently after mixing in ratio of 10 parts base to one part curing agent by weight. It is desirable to set the mixture for 30 minutes before pouring to remove air introduced during mixing. This allows minimum shrinkage, no cure byproducts, and is repairable with good dielectric properties. When liquid components are thoroughly mixed, the mixture cures to a flexible elastomer, which is suited for the protection of PV cell and glass. The glass at the top front should have good transmission in most of the range of solar wavelength, and possess low reflectivity. It should also be impervious to water and be able to take a hit. Tempered glass is often used to withstand all types of weather including heat, cold, and hail storms with a thickness of 1/4″ for strength and durability.

Framing of PV Module

Framing of the PV module is the final step in the three-step process of making solar panels which will provide strength to the entire module. Frames can be made from wood, stainless steel, plastic or aluminum but it should be lightweight and sturdy. Surface-treated aluminum is often used to protect it against corrosion in an adverse weather conditions which also requires much less upkeep. With proper design, aluminum will not have any pits or protrusions for water to gather and enter the module made of 1/8″ aluminum angle in at least 1″ dimension, so that they will last for 30 years or more. Miscellaneous hardware such as 1/4″ machine screws, 1/4″ nuts, 1/4″ lock washers, small angle brackets to join corners of the aluminum frame together, and angle brackets to hold the glass in place are also required. Holes are drilled on the aluminum frame for mounting brackets to securely place them.

The complete sets of a PV module with frame, PV cell, junction box, adhesives, glass, protective film and back sheet laminate and other components is shown in Fig. B.9.

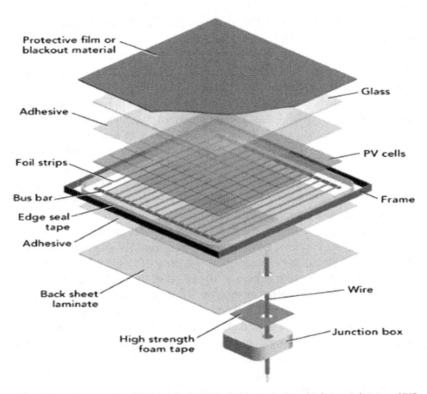

FIGURE B.9 Components of PV module. *Published with permission of Fabrico. A division of EIS.*

Making a Solar-Powered Cell Phone Charger

A solar-powered cell phone charger uses some of the methods employed in the making of the PV module. A solar-powered cell phone charger is a device which can charge cell phones and other mobile devices using sunlight. The major component in this device is a solar panel that converts sunlight into stable voltage for charging cell phones. Solar-powered phone chargers are invaluable for charging cell phones in the remote areas of the world that have no access to electricity. Solar-powered phone chargers can also be used to charge mobile devices in case of power outages due to natural disasters, during camping trips and places with no access to power outlets.

Most of the mobile phones are terminated with a USB cable containing four connection terminals. The connection terminals at the two extreme ends are the supply terminal, which carry 5 V DC in a female USB connector. All of the mobile devices are currently charged with a 5–5.5 V source and current depending upon the capacity of the battery which is typically limited to 500 mA. Solar charging is slow as sunlight falling on the solar panel can vary due to clouds, hence keeping the charger in sunlight may present a challenge. As a result, battery backup solutions are needed. Solar-powered phone chargers can be made as a DIY project with only limited technical knowledge and inexpensive components.

In making one's own solar charger, the important steps to consider are determining the size of the PV cells, making the panel, encapsulating and framing it and terminating the output of the charger with proper cables and plugs. The size of the solar panel is determined by the amount of current and voltage as well as the size of the enclosure. Solar panels should be housed in a manageable enclosure in order to be placed in a backpack, pocket, or a foldable casing with a USB output to charge the mobile devices. Portable solar phone chargers come in different sizes.

One method of making a solar phone charger with an output of 5.0–5.5 V is to use 12 polycrystalline silicon PV cells with an output of 0.5 V connected in series which will work under full sun and low light conditions. This can be achieved by cutting 3 large PV cells measuring 3 × 6 inches with an output of 0.5 V/3.6 A. These 3 large PV cells when split into 4 smaller cells will result into a total of 12 smaller cells. These 12 small PV cells with an output of 0.5 V each connected in series will produce 6 V and 800 mA under full sunlight. A resistor can be used to drop the voltage to the required value. The voltage of the PV cell is independent of the size but the current depends on the size of the PV cell. The constructed solar panel is encapsulated in silicone and is covered with a glass casing within an aluminum frame. A picture frame measuring 9 × 12 inches can be used to enclose the solar charger by gluing the glass to the solar panel with silicone caulk. The original PV cells are cut along the copper ribbon already attached to the cell by using an inexpensive diamond cutting wheel. This design does not require any electronic components to stabilize the system.

The next step after checking the output voltage of the solar charger with a multimeter is to connect a cord with a proper plug that fits your cell phone. Although USB terminals are universally used, there are still many other types of connectors which are being used for different cell phones. One way of solving this problem is to use a cord and plug from an older cell phone. A larger version of the above design is used for kiosk charging systems where multiple cell phones can be charged. Making a kiosk charging system will require larger solar panels with higher current.

This method of making a solar charger is currently used by local inhabitants in a cooperative in the district of Sabana Grande, Nicaragua, and by other small businesses in Africa, South Asia, and South America. Making a solar cellphone charger by this method is inexpensive and is a good educational tool for introducing solar energy as a cottage industry in remote areas of the world. However, the finished product is large in size and is cost ineffective as many PV cells break when mechanically cutting the cells which leads to a large amount of wastage.

This wastage problem was solved by MIT (Massachusetts Institute of Technology, USA) students who developed electronic circuitry as an attachment to step up the voltage of three original cells producing 1.5 V to the required 5 V for charging the local cell phones. The MIT students designed a DC step-up converter and a small solar panel for the charger to obtain the required voltage and power as a part of their final research project at Sabana Grande. This attachment helped to obtain the required 5 V to regulate the current during the charging process of commonly used local cell phones in 10–12 hours of direct sunlight. The attachment eliminates the need to cut the solar cells and also reduces the size of the finished product. The cooperative in Sabana Grande also makes large solar panels at affordable prices for local usage. This cooperative is composed of 20 women and 2 men called the Solar Women of Totogapla (http://www.grupofenix.org/solarwomen/).

The full report of the students' final project is given at (http://ocw.mit.edu/courses/edgerton-center/ec-711-d-lab-energy-spring-2011/projects/MITEC_711S11_proj_rptchrg.pdf) as a part of MIT OpenCourseWare: http://ocw.mit.edu

In United States, more efficient monocrystalline solar panels of 5–6 V output and different sizes are available that can be used to make solar cell phone chargers instead of cutting the PV cells. Some of these suppliers include Amazon.com, eBay, Adafruit, Alibaba, Sundance solar, Voltaic and others that have a good selection of 5–6 V solar panels at affordable prices.

PV Modules in Remote Areas as a Cottage Industry

PV modules can be assembled from sets of small solar cells in remote areas with the help of local people and resources instead of importing them into these locations. Making solar panels from solar cells makes it easier to transport in small units and is less costly as the local workforce is used to assemble them. This can also help local people start small businesses to support projects such as lighting of their homes and schools, powering medical clinics, and using PV modules to generate electricity for pumping and purifying water. Dr. Richard Komp of Skyheat Associates, initiated this work in early 1977 by organizing hands-on workshops on topics such as "Assembling your own solar modules," "Solar charger for cell phones," "Micro-drip irrigation," and "Solar cooker" throughout the world and United States. He uses inexpensive solar cells bought from factory closeouts or used solar cells to teach rural communities about solar projects. Dr. Komp of Skyheat Associates, pioneered "Solar as a cottage industry" in many African, Asian and South American countries. He has taught people to build small PV systems starting with a 2W battery charger that can be used to charge both rechargeable and non-rechargeable batteries such as AA, AAA, C, and D. He also taught them to make bigger systems depending on the need and budget of the community. Making a 5W solar-powered cell phone charger is very popular in many remote communities. This relatively inexpensive and easy to make charging system has helped many people to start small businesses and sustain it after learning from Dr. Komp.

One example is the case of Taurag nomads in the Sahara desert who are making and selling hundreds of small solar-powered cell phone chargers and 20W "Kiosk Model" that can charge up to 5 cell phones at a time in a few minutes. They are also building 16W–18V PV modules to power small PV systems for one room huts in remote villages. All of these systems are built from local components except the imported PV cells. Dr. Komp has also taught people to make big solar ovens to heat cure ethylene acetate as an encapsulant for making 32 and 65W systems for large homes.

In other examples, such as Suni Solar in Nicaragua, bigger systems have been installed—24 solar power cell phone towers for the rain forest region along the Miskito coast. Other groups such as the Grupo Fenix based in Totogalpa, Nicaragua, make their own solar modules for various projects and also offer workshops on how to make such modules. Grupo Fenix works with communities in Nicaragua to promote renewable energy, reforestation and sustainable development.

The cottage industry not only creates income, but it also allows participants to take control of their lives and bring earning into their community. The development of cottage industries is an approach that can boost the rural economy, diversify its production, and alleviate poverty.

Since 2007, Dr. Richard Komp's work on "Solar as a cottage industry" around the world is illustrated in the following 25 trips shown at www.mainesolar.org and www.dadsolar.com. These websites contain reports and photos of the workshops which he conducted at these places.

- Two trips in 2015 to Cameroon (Africa) and Indonesia (South East Asia)
- Two trips in in 2014 to Ghana (Africa) and Miskito coast, (South America)
- One trip in 2013 to Nicaragua (South America)

- Three trips in 2012 to Ghana (Africa), Albuquerque, New Mexico (United States), Northern Honduras (South America)
- Three trips in 2011 to Colombia (South America), Amazon, Brazil (South America), Niger (Africa)
- Four trips in 2010 to Peruvian Andes (South America), Mali (Africa), Nicaragua (South America), and Rwanda (Africa)
- Four trips in 2009 to Rwanda (Africa), Pakistan (South Asia), and twice to Peru (South America)
- Four trips in 2008 to Chile (South America), Peru (South America), Mexico (North America), and Nicaragua (South America)
- Two trips in 2007 to India (South Asia) and Haiti (Central America)

Dr. Richard Komp is a cofounder and currently serves as director of Skyheat Associates, which is recognized as a public nonprofit 509(a) (2) organization part of 501 (c) (3).

Public Solar-Powered Cell Phone Chargers

Using a cell phone is the most effective method of communicating with the world in the aftermath of disasters or in remote areas with no access to the electric grid. There are more than 7 billion active cell phone subscriptions in the world and this is increasing rapidly, especially in the developing countries. Despite this increase, more than 1.3 billion people, mostly in developing countries, do not have access to the electric grid and over 2 billion mobile phone users are without a reliable charging source. In the absence of the electric grid, it is difficult to charge the batteries in a cell phone. Solar PV can be used to charge cell phone batteries which is an alternative to a conventional electrical cell phone charger. Many manufacturers are now marketing solar powered cell chargers in different shapes, sizes, and capacity. A portable charger is good for one cell phone and for one person at a time. Given the increase of mobile phones in the world and the limited charging facilities in remote areas, public parks, street squares, colleges campuses and metropolitans, need is arisen to install large stations with multiple outlets for phones and mobile devices.

According to the European Commission, the first of public solar cell phone charger for mobile devices in the world called the "Strawberry Tree" was developed by a group of students from the University of Belgrade. It was installed in Obrenovac, Serbia, in 2010. The Strawberry tree, as shown in Fig. B.10, is powered by solar panels. This charger produces 900 W of power and has a battery capacity of 4500 Wh. The rechargeable battery accumulates energy from solar panels and can power the charger for more than 14 days without sunshine.

The Strawberry Tree charger contains 16 cords for different type of mobile devices including mobile phones, mp3 players and cameras. It is equipped with 6 USB ports and 8 cables for charging mobile devices and can provide up to 800,000, 10-minute charges annually or 4500, 10-minute charges without sunshine. It also provides a free Wi-Fi service in the surrounding areas. Strawberry Energy Inc.,

FIGURE B.10   Strawberry in Obrenovac, Serbia.

also makes two smaller versions, Strawberry Mini and Mini Rural which are portable and can be carried easily to the location where needed. Strawberry Mini could be used for social events or festivals and Mini Rural, the smallest charger, is designed specifically for remote areas with no access to electricity.

In the United States, AT&T in partnership with Goal Zero and Pensa Design, installed 25 free Street Charge solar-powered charging stations in five boroughs of New York City in 2013. Since then, 15 more locations at parks and beaches have been added. The AT&T Street Charge station is powered with a 15 W monocrystalline solar panels and is capable of charging six devices at a time. It can also charge small electronic devices via the USB cables. AT&T Street Charge, initially started as a direct outgrowth of Hurricane Sandy, will continue to offer charging facilities in 2016. There are 40 charging stations in more than 25 parks, beaches, and outdoor locations throughout 5 boroughs of New York city.

In 2013, WrightGrid in United States developed and installed a solar-powered public charging system, Model Z, in Boston, Massachusetts, as shown in Fig. B.11. The system consists of 150 W solar panel which produces 5.25 V/2.5 A from each of 10 secure cell phone lockers. It has a battery backup of 210 Ah. The cell phone can be charged free of cost with universal cables that can charge up to 95% of the mobile and tablet devices including all Apple's iOS (originally iPhone OS) and Android devices.

Model Z is a solar-powered system that can be used to charge mobile devices in the off-grid remote areas of the world as well as college campuses, trade shows, recreation places, and outdoor festivals where no public power source outlet is available. The system, as shown in Fig. B.11, measures 100 inches in height, 56 inches in width (including ad panel) and 23 inches in width. It weighs 550 lbs, withstands wind speeds of 60 miles/hour and has optimal operating temperature from 20°F to 105°F.

FIGURE B.11   WrightGrid Model Z public phone charging system, Boston, United States.

FIGURE B.12   Palm tree public phone charging system in Dubai, UAE.

In 2015, Dubai installed two solar-powered palm trees to provide eight charging ports for cell phones and Wi-Fi hot spots with a broadcast range of 200 m (328 feet) which can connect up to 50 users at one time. The palm tree as shown in Fig. B.12 is 6 m tall and consists of nine leaf-shaped photovoltaic modules which produce around 7.2 kWh/day of electricity to power the smart palm.

APPENDIX C. RESULTS OF SOLAR INSOLATION/RADIATION USING DIFFERENT DATABASES

This appendix presents the results and methods of calculating solar isolation/ radiation using the following databases:

- NASA SSE 6.0
- RETScreen International
- SolarGISiMaps and SolarGIS pvPlanner
- NREL Renewable Resource Data Center

Calculation of Solar Insolation/Radiation Using NASA SSE 6.0

The NASA SSE 6.0 website uses a friendly interface to access data on the fly by entering the latitude and longitude of a location of the user's interest. The latitude and longitude of the specified place of interest can be found from the website http://www.latlong.net/convert-address-to-lat-long.html. By 2015, the SSE 6.0 website has grown to contain more than 200 satellite-derived meteorology and solar energy parameters of monthly averages from 22 years of data. A condensed list of these parameters with definitions is given in Table III.2 Power/SSE Release 6.0 Archive: Parameters, Temporal coverage and Data sources, which can be accessed at the following website.

http://power.larc.nasa.gov/common/SSE_Methodology/Tables/SSE_ MethodologyTableIII2.html

Users can select the parameters of their interest and the selections which are grouped by their most probable applications. The following examples illustrates the method of calculating the solar insolation/radiation using SSE 6.0.

1. Go to the NASA website https://eosweb.larc.nasa.gov/cgi-bin/sse/grid.cgi.
   This will take you directly to the NASA Meteorological and Solar Energy area where latitude and longitude for the interested location can be entered.
2. Click Submit after entering latitude/longitude of the place of interest. This will take you to the NASA Surface meteorology and Solar energy choices.
   a. Choose: Parameters for sizing and pointing of solar panels as Insolation on horizontal surface (average, Min and Max)
   b. Chose: Parameters for Tilted Solar Panels as Radiation on equator points tilted surfaces
   c. Chose: Parameter for Meteorology(temperature) as Air temperature at 10 m
   d. Choose: Parameter for Meteorology (wind) as Wind speed at 50 m
3. After selecting the parameters click Submit. This will take you to the NASA Surface meteorology and Solar Energy—Available tables (results) as shown in Figs. C.1 and C.2.

NASA Surface meteorology and Solar
Energy -   Available Tables

ATMOSPHERIC SCIENCE DATA CENTER

Latitude 23.81 / Longitude 90.412 was chosen

Geometry Information   Elevation: 50 meters taken from the NASA GEOS-4 model
elevation

Northern boundary
24

Western boundary	Center	Eastern boundary
90	Latitude 23.5	91
	Longitude 90.5	

Southern boundary   23

Parameters for Sizing and Pointing of Solar Panels and for Solar Thermal Applications:
Monthly Averaged Insolation Incident On A Horizontal Surface (kWh/m²/day)

Lat 23.81 Lon 90.412	Jan	Feb	Mar	Apr	May	Jun	Jul	Aug	Sep	Oct	Nov	Dec	Annual Average
22-year Average	4.36	4.92	5.59	5.76	5.30	4.53	4.23	4.29	4.01	4.32	4.28	4.21	4.64

Minimum And Maximum Difference From Monthly Averaged Insolation (%)

Lat 23.81 Lon 90.412	Jan	Feb	Mar	Apr	May	Jun	Jul	Aug	Sep	Oct	Nov	Dec
Minimum	-21	-11	-16	-10	-12	-14	-17	-14	-19	-15	-13	-17
Maximum	13	7	9	10	16	16	10	16	21	13	10	9

Meteorology (Temperature):
Monthly Averaged Air Temperature At 10 m Above The Surface Of The Earth (°C)

Lat 23.81 Lon 90.412	Jan	Feb	Mar	Apr	May	Jun	Jul	Aug	Sep	Oct	Nov	Dec	Annual Average
22-year Average	19.7	23.0	26.4	27.1	27.6	27.9	27.6	27.6	27.0	25.4	22.5	20.2	25.2
Minimum	14.5	17.9	21.7	23.5	24.7	25.6	25.4	25.3	24.5	22.4	18.6	15.4	21.6
Maximum	24.9	27.6	30.7	30.7	30.5	30.2	29.8	29.9	29.7	29.3	27.5	25.7	28.9

Meteorology (Wind):
Monthly Averaged Wind Speed At 50 m Above The Surface Of The Earth (m/s)

Lat 23.81 Lon 90.412	Jan	Feb	Mar	Apr	May	Jun	Jul	Aug	Sep	Oct	Nov	Dec	Annual Average
10-year Average	2.36	2.66	2.83	3.19	3.20	3.07	2.74	2.41	2.17	1.85	2.08	2.20	2.56

Minimum And Maximum Difference From Monthly Averaged Wind Speed At 50 m (%)

Lat 23.81 Lon 90.412	Jan	Feb	Mar	Apr	May	Jun	Jul	Aug	Sep	Oct	Nov	Dec	Annual Average
Minimum	-11	-16	-19	-13	-13	-11	-8	-9	-9	-11	-13	-12	-12

FIGURE C.1   Result of three selected parameters for a location near Dhaka, Bangladesh.

Fig. C.1 shows parameters such as monthly average insolation, monthly air temperature and monthly wind speed for Dhaka in Bangladesh.

Fig. C.2 shows monthly solar insolation at different tilt angles for the same city Dhaka in Bangladesh.

 ATMOSPHERIC SCIENCE DATA CENTER   NASA Surface meteorology and Solar Energy - Available Tables Latitude 23.81 / Longitude 90.412 was chosen.

Geometry Information Elevation: 50 meters taken from the NASA GEOS-4 model elevation

Northern boundary
24

Western boundary   Center   Eastern boundary
90   Latitude 23.5   91
Longitude 90.5

Parameters for Sizing and Pointing of Solar Panels and for Solar Thermal Applications:

Monthly Averaged Insolation Incident On A Horizontal Surface (kWh/m²/day)

Lat 23.81 Lon 90.412	Jan	Feb	Mar	Apr	May	Jun	Jul	Aug	Sep	Oct	Nov	Dec	Annual Average
22-year Average	4.36	4.92	5.59	5.76	5.30	4.53	4.23	4.29	4.01	4.32	4.28	4.21	4.64

Minimum And Maximum Difference From Monthly Averaged Insolation (%)

Lat 23.81 Lon 90.412	Jan	Feb	Mar	Apr	May	Jun	Jul	Aug	Sep	Oct	Nov	Dec
Minimum	−21	−11	−16	−10	−12	−14	−17	−14	−19	−15	−13	−17
Maximum	13	7	9	10	16	16	10	16	21	13	10	9

Parameters for Tilted Solar Panels:

Monthly Averaged Radiation Incident On An Equator-Pointed Tilted Surface (kWh/m²/day)

Lat 23.81 Lon 90.412	Jan	Feb	Mar	Apr	May	Jun	Jul	Aug	Sep	Oct	Nov	Dec	Annual Average
SSE HRZ	4.36	4.92	5.59	5.76	5.30	4.53	4.23	4.29	4.01	4.32	4.28	4.21	4.64
K	0.62	0.60	0.59	0.54	0.48	0.40	0.38	0.40	0.41	0.50	0.58	0.63	0.51
Diffuse	1.07	1.37	1.72	2.12	2.42	2.50	2.44	2.32	2.08	1.69	1.23	0.97	1.83
Direct	6.44	6.20	6.08	5.34	4.13	2.91	2.56	2.84	2.94	4.42	5.79	6.59	4.68
Tilt 0	4.23	4.88	5.52	5.65	5.27	4.51	4.21	4.19	3.96	4.28	4.19	4.10	4.58
Tilt 8	4.71	5.27	5.76	5.71	5.22	4.44	4.16	4.19	4.04	4.52	4.62	4.62	4.77
Tilt 23	5.40	5.78	5.96	5.60	4.94	4.18	3.94	4.06	4.04	4.79	5.22	5.39	4.94
Tilt 38	5.79	5.98	5.85	5.22	4.44	3.74	3.57	3.76	3.87	4.83	5.54	5.86	4.86
Tilt 90	4.62	4.23	3.36	2.34	1.86	1.70	1.67	1.80	2.15	3.19	4.30	4.85	3.00
OPT	5.87	5.98	5.96	5.71	5.27	4.51	4.21	4.20	4.06	4.84	5.58	5.99	5.18
OPT ANG	48.0	40.0	25.0	10.0	0.00	0.00	0.00	4.00	16.0	33.0	46.0	51.0	22.6

NOTE: *Diffuse radiation, direct normal radiation and tilted surface radiation are not calculated when the clearness index (K) is below 0.3 or above 0.8.*

FIGURE C.2   Results of selected parameters at different tilt angle for a location near Dhaka, Bangladesh.

Calculation of Solar Insolation Using RETScreen International

RETScreen also uses the SSE 6.0 database which can be accessed as given below. The results of the following fixed parameters will be obtained unlike the above example where there is a choice of selecting different parameters.

NASA Surface meteorology and Solar
Energy: RETScreen Data

Latitude 31.4 / Longitude 121.5 was chosen.

	Unit	Climate data location
Latitude	°N	31.4
Longitude	°E	121.5
Elevation	m	12
Heating design temperature	°C	0.17
Cooling design temperature	°C	28.91
Earth temperature amplitude	°C	14.44
Frost days at site	day	13

Month	Air temperature	Relative humidity	Daily solar radiation - horizontal	Atmospheric pressure	Wind speed	Earth temperature	Heating degree-days	Cooling degree-days
	°C	%	kWh/m²/d	kPa	m/s	°C	°C-d	°C-d
January	4.8	73.7%	2.63	102.5	4.9	5.3	404	2
February	5.9	73.5%	3.09	102.3	4.9	6.3	340	6
March	9.0	74.9%	3.53	102.0	4.5	9.5	278	26
April	14.0	77.0%	4.28	101.5	4.3	14.4	130	121
May	18.5	79.8%	4.84	101.0	4.0	18.9	27	262
June	22.4	84.6%	4.58	100.5	4.1	23.0	0	376
July	25.9	87.2%	5.17	100.4	4.2	26.6	0	501
August	25.8	86.5%	4.78	100.5	4.1	26.6	0	495
September	22.7	80.8%	4.07	101.1	4.3	23.1	0	383
October	17.9	75.1%	3.36	101.8	4.3	18.3	33	245
November	12.7	73.9%	2.79	102.2	4.6	13.2	158	97
December	7.3	73.5%	2.58	102.6	4.7	7.8	326	14
Annual	15.6	78.4%	3.81	101.5	4.4	16.1	1696	2528
Measured at (m)					10.0	0.0		

FIGURE C.3 Results of fixed parameters for RTEScreen data at a location near Shanghai, China.

1. Go to https://eosweb.larc.nasa.gov/sse/RETScreen/.
   This will take you to the NASA Surface meteorology and Solar energy: RETScren Data.
2. Enter both the latitude and longitude either in decimal degrees or degrees and minutes separated by space. The results for the following fixed parameters for Shanghai, China are given in Fig. C.3:
   - Air temperature
   - Relative humidity
   - Daily solar radiation–horizontal

- Atmospheric pressure
- Wind speed
- Earth temperature
- Heating degree days
- Cooling degree days

Estimation of Solar Insolation for International Locations Using SolarGIS iMaps

SolarGIS iMaps is an interactive map that provides global solar radiation and meteorological information. These solar radiation maps shows detailed and accurate information of global horizontal irradiation (GHI) and direct normal irradiation (DNI) at a spatial resolution of 90 m from the chosen locations worldwide. It also provides information about diffused horizontal irradiation, temperature, terrain, and population data. Long-term monthly averages can be downloaded as charts and tables with PDF, XLS and CVS formats. A free trial version of iMaps, solar radiation maps and solar resource maps of various countries and regions are also available. The following examples illustrate the method of calculation of solar insolation/radiation.

Go to website: http://solargis.info/doc/about-imaps.

This will take you to the world map on the right hand side of the screen as shown in Figs. C.4 and C.5 with the following information on the left hand side of screen.

- Search map
- Three windows consisting of "Name of location or latitude/longitude of the location," a window for "search" and another window for "clear" will

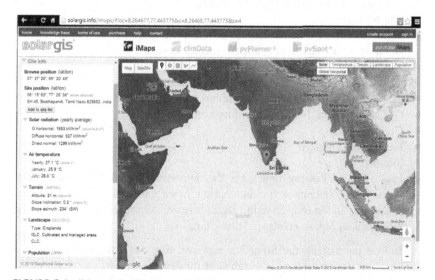

FIGURE C.4   iMap result of Boothapandi, India.

FIGURE C.5   iMap result of Geraldton, Australia.

appear. When entering the name of the location or latitude/longitude coordinates and clicking search, this will take you to the actual position on the map. It will also provide all the information except solar radiation (yearly average) which is a paid access.

- Site info: Browse position (latitude/longitude)
- Site position
- Solar radiation (yearly average): G horizontal (paid access), Diffuse horizontal (paid access), Direct normal (paid access)
- Air temperature: Yearly, Jan., Jul.
- Terrain: Altitude, Slope inclination, Slope azimuth
- Landscape: Type: Deciduous, continuous broad leaves, forests
- GLC: Tree cover, broaded, deciduous, close
- CLC: Airports
- Population: Density
- Site list
- List of the following nine sites for demonstration: Boothapandi, India; Aosta, Italy; Stellenbosch, South Africa; Arica, Chile; Geraldton, Australia; Sukapura, Indonesia; Riyadh, Saudi Arabia; San Diego, United States; Fujivoshida, Japan, Zhangype, Shi, China.

Before purchasing the SolarGIS services, one can download iMap as a free trial version from the list of ten sites for demonstration. The result of the first iMap location is in Boothapandi, India which is north of the equator as shown in Fig. C.4. The second location of the iMap is in Geraldton, Western

Australia which is south of the equator as shown in Fig. C.5. Both of these locations are situated near the coastal areas that are prone to natural disasters.

Estimation of Electric Potential and Solar Radiation for International Locations Using SolarGIS pvPlanner

SolarGIS pvPlanner is an online tool which can calculate the electric potential of a photovoltaic system in minutes for any selected location. This is a useful tool which can be used in conjunction with iMaps for designing and analyzing an efficient PV system. SolarGIS pvPlanner can also be used for comparing the power yield from different PV technologies such as crystalline silicon (c-Si), amorphous silicon (a-Si), cadmium telluride (cdTe), copper indium selenide (CIS) as well as benefits of fixed, and 1-axis or 2-axis tracking PV systems. The tool's simulation uses and works with the most accurate solar radiation, high resolution climatic and geographic data. The results are displayed by means of monthly values and daily profiles of solar radiation and photovoltaic power output both for tilted and sun-tracking PV systems. The results can be downloaded in PDF, XLS, and CSV formats with the option of multiple languages, and is available free of cost for trial usage.

The calculation of photovoltaic electricity potential using SolarGIS pvPlanner can be performed in the following three steps.

1. Go to http://solargis.info/pvplanner/#tl=Google:roadmap&bm=map
   a. This will take you to the world map on the right hand side of the screen and three set of information about the location of interest on the left hand side of the screen. This is shown in Figs. C.6 and C.7.
   b. Enter the address or coordinates (lat/long) of interest. The coordinates of locations can also be found from the interactive search map.
2. Click on the continue button for proceeding to the next step. This will take you to the next screen which require the following PV system configuration details to be entered:
   a. Power installed in kWp
   b. Module type (c-Si or a-Si)
   c. Inverter efficiency
   d. Module tilt and azimuth
   e. Mounting type (fixed axis or tracker system)
   f. Optional parameter to modify the horizon to account for shading losses from nearby objects such as trees and building is also provided.
3. Click on the Calculate button after entering these values. This will take you to the next screen with the following results as shown in Figs. C.8 and C.9.
   a. PV electricity potential for each month and annual in the following units
   Etm: Monthly sum of total electricity prod (KWh)
   Esm: Monthly sum of specific electricity prod (KWh/kWp)
   Esd: Daily sum of specific electricity prod (KWh/kWp)
   Eshare: Percentual share of monthly electricity prod (%).
   PR: Performance ratio (%)

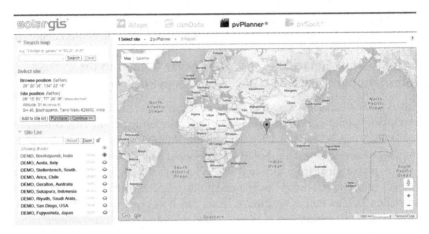

FIGURE C.6   pvPlanner for Boothapandi, India.

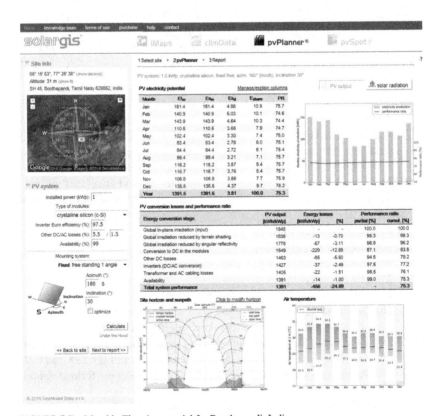

FIGURE C.7   Monthly Electric potential for Boothapandi, India.

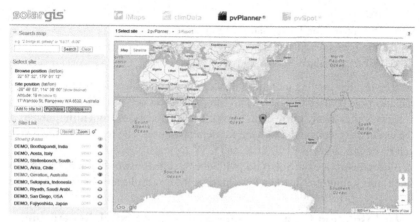

FIGURE C.8 pvPlanner result of Geraldton, Australia.

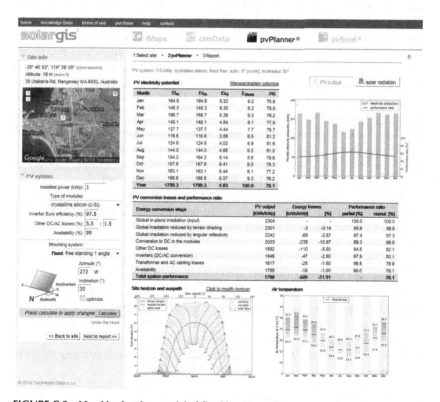

FIGURE C.9 Monthly electric potential of Geraldton, Australia.

b. PV output

c. PV conversion losses and performance ratio

d. Site horizon and sunpath

e. Air temperature

The results showing monthly statistics of potential electrical output and solar radiation can be used to analyze and compare different types of PV technologies at various sites at any place in the world. Ten site locations from different part of the world are given on the website for demonstration. The reports for the results can be downloaded in PDF, XLS, and CSV formats.

Given below are the SolarGIS pvPlanner results for two sites at Boothapandi, India and Geraldton, Australia for which the iMaps results are already calculated.

Solar Insolation Data for United States Using NREL Renewable Resource Data Center

The NREL, through its Renewable Resource Data Center (RReDC), provides free online assessments of solar, wind and other energy resources in the United States. The RReDC's Solar Radiation Data Manual for Flat-Plate and Concentrating Collectors provides monthly average, minimum, and maximum values from data collected from 239 sites throughout the United States. The data accounts for variations in site and weather conditions at various times of the year and south-facing array orientations over the 30-year period from 1961 through 1990. The files containing individual PDFs for the manual and site data tables can be downloaded in three compression formats: PC, UNIX and Macintosh. Derived maps of the data in the tables are also available for viewing. The Solar Radiation Data Manual for Flat-Plate and Concentrating Collectors can be found at http://www.nrel.gov/docs/legosti/old/5607.pdf.

The updated National Solar Radiation Database (http://rredc.nrel.gov/solar/old_data/nsrdb/1991-2010/) contain solar and meteorological data from 1991 to 2010 for 1454 sites and is available on the internet for United States and its territories.

Each data sheet shown in Figs. C.10 and C.11 includes important information about the selected site, such as elevation, latitude and longitude, average climate condition, and barometric pressure. The latitude of the site is used to determine array orientation and barometric pressure which may be used in computing the corrected air mass value at the site. The solar resource information is arranged in tables and columns. The first four tables contains solar radiation or peak hours in units of kWh/m^2 per day for the following four systems and fifth table relates to average climatic conditions as shown below:

1. South-facing fixed-tilt collectors

2. Single axis tracking collector

3. Dual axis tracking collectors

4. Direct-beam concentrating collectors

5. Average climatic conditions

The fixed and single axis tracking collector tables are broken down into tilts of horizontal, latitude minus 15 degrees, latitude, and latitude plus 15 degrees. The yearly average annual average solar insolation is used for PV system sizing and performance analysis while the data on south facing fixed-tilt collectors are used in most PV system installations.

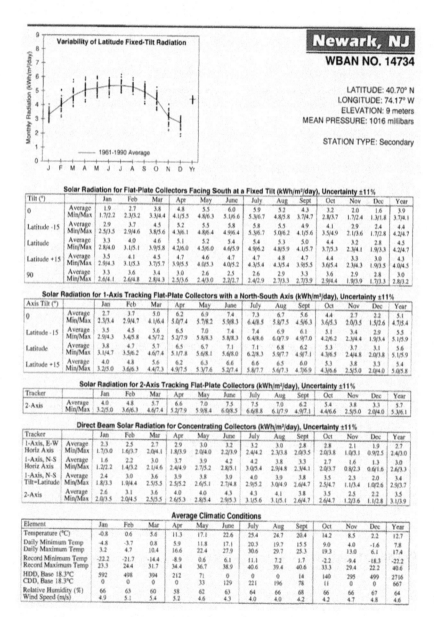

FIGURE C.10   Data set for Newark, New Jersey.

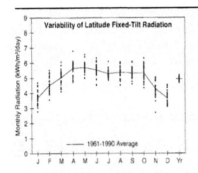

New Orleans, LA

WBAN NO. 12916

LATITUDE: 29.98° N
LONGITUDE: 90.25° W
ELEVATION: 3 meters
MEAN PRESSURE: 1017 millibars

STATION TYPE: Secondary

Solar Radiation for Flat-Plate Collectors Facing South at a Fixed Tilt (kWh/m²/day), Uncertainty ±9%

Tilt (°)		Jan	Feb	Mar	Apr	May	June	July	Aug	Sept	Oct	Nov	Dec	Year
0	Average	2.7	3.6	4.5	5.5	6.1	6.1	5.7	5.5	4.9	4.3	3.1	2.6	4.6
	Min/Max	2.3/3.2	3.0/4.2	3.8/5.2	4.7/6.6	5.3/7.0	5.0/7.1	5.3/6.6	4.7/6.1	4.3/5.5	3.7/5.0	2.3/3.6	2.3/3.0	4.3/4.8
Latitude -15	Average	3.3	4.2	4.9	5.7	6.0	6.0	5.7	5.6	5.3	5.0	3.8	3.2	4.9
	Min/Max	2.6/4.1	3.3/5.0	4.0/5.8	4.8/6.9	5.3/7.0	4.8/6.9	5.2/6.5	4.7/6.3	4.6/6.0	4.1/5.8	2.5/4.4	2.8/3.9	4.6/5.1
Latitude	Average	3.7	4.5	5.0	5.6	5.7	5.5	5.3	5.4	5.3	5.3	4.3	3.7	5.0
	Min/Max	2.8/4.7	3.5/5.5	4.1/6.1	4.7/6.8	5.0/6.6	4.5/6.3	4.8/6.1	4.5/6.1	4.6/6.1	4.3/6.3	2.7/5.1	3.1/4.6	4.7/5.2
Latitude +15	Average	3.9	4.6	4.9	5.3	5.1	4.8	4.7	4.9	5.1	5.4	4.5	3.9	4.8
	Min/Max	2.8/5.1	3.6/5.7	4.0/6.0	4.4/6.3	4.4/5.8	4.0/5.4	4.3/5.3	4.1/5.5	4.4/5.9	4.3/6.5	2.7/5.4	3.3/4.9	4.5/5.0
90	Average	3.3	3.5	3.2	2.7	2.1	1.8	1.9	2.3	3.0	3.9	3.7	3.4	2.9
	Min/Max	2.3/4.4	2.7/4.4	2.6/3.9	2.3/3.0	1.9/2.2	1.7/1.9	1.8/2.0	2.1/2.5	2.7/3.5	3.0/4.7	2.1/4.5	2.7/4.4	2.7/3.0

Solar Radiation for 1-Axis Tracking Flat-Plate Collectors with a North-South Axis (kWh/m²/day), Uncertainty ±9%

Axis Tilt (°)		Jan	Feb	Mar	Apr	May	June	July	Aug	Sept	Oct	Nov	Dec	Year
0	Average	3.6	4.7	5.8	7.1	7.7	7.6	7.0	6.8	6.3	5.8	4.2	3.5	5.8
	Min/Max	2.7/4.6	3.6/6.0	4.5/7.2	5.7/9.1	6.5/9.3	5.8/9.3	6.3/8.3	5.5/7.9	5.2/7.4	4.6/7.1	2.6/5.0	2.8/4.3	5.5/6.2
Latitude -15	Average	4.0	5.2	6.1	7.3	7.7	7.5	6.9	6.9	6.6	6.3	4.7	3.9	6.1
	Min/Max	3.0/5.2	3.9/6.6	4.8/7.6	5.8/9.3	6.5/9.3	5.8/9.1	6.2/8.2	5.5/8.0	5.4/7.7	4.9/7.8	2.8/5.7	3.2/5.0	5.7/6.5
Latitude	Average	4.3	5.5	6.3	7.3	7.4	7.2	6.7	6.7	6.6	6.6	5.1	4.3	6.2
	Min/Max	3.1/5.7	4.1/7.0	4.8/7.8	5.8/9.3	6.3/9.0	5.5/8.8	6.0/7.9	5.4/7.9	5.4/7.8	5.1/8.1	2.9/6.1	3.4/5.5	5.9/6.6
Latitude +15	Average	4.5	5.6	6.2	7.0	7.0	6.7	6.2	6.4	6.5	6.7	5.3	4.5	6.0
	Min/Max	3.2/6.0	4.1/7.1	4.7/7.8	5.5/8.9	5.9/8.5	5.1/8.1	5.6/7.4	5.1/7.5	5.3/7.6	5.1/8.2	2.9/6.4	3.6/5.8	5.6/6.4

Solar Radiation for 2-Axis Tracking Flat-Plate Collectors (kWh/m²/day), Uncertainty ±9%

Tracker		Jan	Feb	Mar	Apr	May	June	July	Aug	Sept	Oct	Nov	Dec	Year
2-Axis	Average	4.5	5.6	6.3	7.3	7.8	7.7	7.0	6.9	6.6	6.7	5.3	4.5	6.4
	Min/Max	3.2/6.1	4.2/7.2	4.8/7.9	5.9/9.4	6.6/9.4	5.8/9.4	6.3/8.4	5.5/8.1	5.4/7.8	5.2/8.2	2.9/6.4	3.6/5.9	5.9/6.8

Direct Beam Solar Radiation for Concentrating Collectors (kWh/m²/day), Uncertainty ±8%

Tracker		Jan	Feb	Mar	Apr	May	June	July	Aug	Sept	Oct	Nov	Dec	Year
1-Axis, E-W Horiz Axis	Average	2.4	2.9	3.0	3.5	3.6	3.5	3.0	3.0	3.0	3.6	2.9	2.6	3.1
	Min/Max	1.2/3.8	1.8/4.2	1.9/4.2	2.2/5.1	2.7/5.0	2.0/4.9	2.4/4.2	1.8/3.9	2.1/4.0	2.4/4.9	1.0/4.0	1.8/3.7	2.8/3.5
1-Axis, N-S Horiz Axis	Average	2.1	3.0	3.6	4.5	4.8	4.5	3.9	3.9	3.8	3.9	2.7	2.1	3.6
	Min/Max	1.1/3.3	1.9/4.4	2.2/5.1	2.9/6.8	3.5/6.7	2.6/6.3	3.1/5.3	2.3/5.1	2.5/5.0	2.6/5.4	0.9/3.7	1.4/3.1	3.1/4.0
1-Axis, N-S Tilt=Latitude	Average	2.7	3.6	4.0	4.6	4.6	4.2	3.6	3.8	4.0	4.5	3.4	2.8	3.8
	Min/Max	1.4/4.2	2.3/5.3	2.5/5.6	3.0/6.9	3.4/6.4	2.4/5.9	2.9/5.0	2.3/5.1	2.7/5.3	3.0/6.3	1.1/4.6	1.9/4.1	3.4/4.3
2-Axis	Average	2.9	3.7	4.0	4.7	4.8	4.5	3.9	3.9	4.1	4.6	3.6	3.0	4.0
	Min/Max	1.5/4.6	2.3/5.4	2.5/5.7	3.0/7.1	3.6/6.7	2.6/6.4	3.2/5.3	2.4/5.2	2.7/5.3	3.0/6.3	1.2/4.9	2.0/4.4	3.5/4.5

Average Climatic Conditions

Element	Jan	Feb	Mar	Apr	May	June	July	Aug	Sept	Oct	Nov	Dec	Year
Temperature (°C)	10.7	12.4	16.4	20.3	23.8	26.7	27.7	27.5	25.6	20.6	16.2	12.5	20.1
Daily Minimum Temp	5.4	6.9	10.9	14.7	18.4	21.6	22.8	22.7	20.8	14.8	10.6	7.1	14.7
Daily Maximum Temp	16.0	17.8	22.0	25.8	29.1	31.8	32.6	32.3	30.3	26.3	21.7	17.9	25.3
Record Minimum Temp	-10.0	-7.2	-3.9	0.0	10.0	10.0	15.6	15.6	5.6	1.7	-4.4	-11.7	-11.7
Record Maximum Temp	28.3	29.4	31.7	33.3	35.6	37.8	38.3	38.9	38.3	33.3	30.6	28.9	38.9
HDD, Base 18.3°C	250	176	90	16	0	0	0	0	0	17	99	194	841
CDD, Base 18.3°C	14	9	31	74	169	250	291	284	218	87	34	13	1475
Relative Humidity (%)	76	73	73	73	74	76	79	79	78	75	77	77	76
Wind Speed (m/s)	4.0	4.3	4.2	4.1	3.6	3.0	2.6	2.6	3.1	3.3	3.8	4.0	3.5

FIGURE C.11   Dataset for New Orleans, Louisiana.

APPENDIX D. STANDARD FOR PV SYSTEMS

(Reproduced from USAID: Powering Health: Electrification option for developing countries health facilities: Photovoltaic (PV) systems, available at http://www.poweringhealth.org/index.php/topics/technology/solar-pv#standards).

There are numerous national and international bodies that set Standards for photovoltaics. There are Standards for nearly every stage of the PV lifecycle, including materials and processes used in the production of PV panels, testing methodologies, performance standards, and design and installation guidelines. The Standards shown below are not a complete list, but are those most relevant to the procurement and installation of solar PV systems. Each standard has been loosely categorized based on its subject matter.

International Electrotechnical Commission (IEC)

Category	Standard
Characteristics	IEC 61194 ed1.0: Characteristic parameters of stand-alone photovoltaic (PV) systems
Crystalline	IEC 61215 ed 2.0: Crystalline silicon terrestrial photovoltaic (PV) modules—Design qualification and type approval
Thin-film	IEC 61646 ed 2.0: Thin-film terrestrial photovoltaic (PV) modules—Design qualification and type approval
Test	IEC 61701 ed 2.0: Salt mist corrosion testing of photovoltaic (PV) modules
Characteristics	IEC 61702 ed1.0: Rating of direct coupled photovoltaic (PV) pumping systems
Monitoring	IEC 61724 ed1.0: Photovoltaic system performance monitoring—Guidelines for measurement, data exchange and analysis
Characteristics	IEC 61727 ed 2.0: Photovoltaic (PV) systems—Characteristics of the utility interface
Safety	IEC 61730-1 ed 1.0: Photovoltaic (PV) module safety qualification—Part 1: Requirements for construction
Safety	IEC 61730-2 ed 1.0: Photovoltaic (PV) module safety qualification—Part 2: Requirements for testing

Category	Standard
Terms	IEC/TS 61836 ed 2.0: Solar photovoltaic energy systems—Terms, definitions and symbols
Balance of System	IEC 62093 ed 1.0: Balance-of-system components for photovoltaic systems—Design qualification natural environments
Balance of System	IEC 62109-1 ed 1.0: Safety of power converters for use in photovoltaic power systems—Part 1: General requirements
Balance of System	IEC 62109-2 ed 1.0: Safety of power converters for use in photovoltaic power systems—Part 2: Particular requirements for inverters
Test	IEC 62116 ed 1.0: Test procedure of islanding prevention measures for utility-interconnected photovoltaic inverters
Design	IEC 62124 ed 1.0: Photovoltaic (PV) stand alone systems—Design verification
Design	IEC 62253 ed 1.0: Photovoltaic pumping systems—Design qualification and performance measurements
Rural electrification	IEC/TS 62257 ed 1.0: Recommendations for small renewable energy and hybrid systems for rural electrification—Parts 1–9
Commissioning	IEC 62446 ed 1.0: Grid connected photovoltaic systems—Minimum requirements for system documentation, commissioning tests and inspection
Performance	IEC 62509 ed 1.0: Battery charge controllers for photovoltaic systems—Performance and functioning
Rural electrification	IEC/PAS 62111 ed 1.0: Specifications for the use of renewable energies in rural decentralised electrification
Balance of System	IEC 60269-6 ed 1.0: Low-voltage fuses—Part 6: Supplementary requirements for fuse-links for the protection of solar photovoltaic energy systems

Category	Standard
Installation	IEC 60364-1 ed 5.0: Low-voltage electrical installations—Part 1: Fundamental principles, assessment of general characteristics, definitions
Installation	IEC 60364-7-712 ed 1.0: Electrical installations of buildings—Part 7-712: Requirements for special installations or locations—Solar photovoltaic (PV) power supply systems

Institute of Electrical and Electronics Engineers (IEEE)

Category	Standard
Performance	IEEE 1526-2003: IEEE Recommended Practice for Testing the Performance of Stand-Alone Photovoltaic Systems
Sizing	IEEE 1562-2007: IEEE Guide for Array and Battery Sizing in Stand-Alone Photovoltaic (PV) Systems
Interconnection	IEEE 1547.2-2008: IEEE Application Guide for IEEE Std 1547(TM), IEEE Standard for Interconnecting Distributed Resources with Electric Power Systems
Interconnection	IEEE 1547.3-2007: IEEE Guide for Monitoring, Information Exchange, and Control of Distributed Resources Interconnected with Electric Power Systems
Interconnection	IEEE 1547.1-2003: IEEE Standard for Interconnecting Distributed Resources with Electric Power Systems

Underwriters Laboratory (UL)

Category	Standard
Crystalline	UL 1703: Standard for Flat-Plate Photovoltaic Modules and Panels
Concentrated	UL 8703: Concentrator photovoltaic modules and assemblies

Category	Standard
Mounting	UL 790: Standard for Standard Test Methods for Fire Tests of Roof Coverings
Mounting	UL 1897: Standard for Uplift Tests for Roof Covering Systems
Balance of System	UL-SU 2703: Rack mounting systems and clamping devices for flat-plate photovoltaic modules and panels
Balance of System	UL 1741: Standard for Inverters, Converters, Controllers and Interconnection System Equipment for Use With Distributed Energy Resources
Balance of System	UL-SU 1699B: Photovoltaic (PV) DC arc-fault circuit protection
Balance of System	UL-SU 4703: Photovoltaic wire
Balance of System	UL 854: Standard for Service—Entrance Cables
Balance of System	UL-SU 2579: Low-voltage fuses—fuses for photovoltaic systems
Balance of System	UL 4248-18: Fuseholders—Part 18: Photovoltaic
Balance of System	UL-SU 6703: Connectors for use in photovoltaic systems
Balance of System	UL-SU 6703A: Multi-pole connectors for use in photovoltaic systems
Test	UL-SU 5703: Determination of the maximum operating temperature rating of photovoltaic (PV) backsheet materials
Balance of System	UL 3730: Photovoltaic junction boxes
Balance of System	UL-SU 98B: Enclosed and dead-front switches for use in photovoltaic systems
Balance of System	UL 489B: Molded-case circuit breakers, molded-case switches, and circuit-breaker enclosures for use with photovoltaic (PV) systems

American Society for Testing and Materials (ASTM)

Category	Standard
Terms	ASTM E772—11: Standard Terminology of Solar Energy Conversion
Test	ASTM E2848—11: Standard Test Method for Reporting Photovoltaic Non Concentrator System Performance

BIBLIOGRAPHY

[1] G.A. Boschloo, A. Hagfeldt, Characteristics of the Iodide/Triiodide Redox Mediator in Dye-Sensitized Solar Cells, American Chemical Society 2009. <http://pubs.acs.org/doi/abs/10.1021/ar900138m>.

[2] Build Solar Panels: DIY Solar Power, Homemade Solar Panels 4 the Average Person. <http://mysearchforsolarpower.com/how-to-build-solar-panels-cost-effectively/>.

[3] A.W. Czanderna, F.J. Pern, Encapsulation of PV modules using ethylene vinyl acetate copolymer as a pottant: A critical review. Measurement and characterization Branch, National Renewable Energy Laboratory, Golden, CO.

[4] Gmoke, Solar as a Cottage Industry. <http://www.dailykos.com/story/2013/03/26/1196968/-Solar-as-a-Cottage-Industry>, 2013.

[5] M. Gratzel, Solar energy conversion by dye-sensitized photovoltaic cells: Inorganic Chemistry, American Chemical Society 2005. <http://pubs.acs.org/doi/abs/10.1021/ic0508371>.

[6] S. Hanley, Enabling Breakthroughs in Solar Technology, Solar Love 2015. Solar love <http://solarlove.org/enabling-breakthroughs-solar-technology/>.

[7] Hands-on Science Kits and Demos: "Nanocrystalline Solar Cell Kit". ICE (Institute for chemical Education, Maidson WI) <http://ice.chem.wisc.edu/Catalog/SciKits.html>.

[8] R. Hertzberg, Solar power for mobile phones in emerging markets. Published in Africa and the Middle East 2008 Industries Related: ENERGY (2015).

[9] N. Jewell, Solar-Powered Smart Palms Provide Beachgoers with Wi-Fi and Charging Stations. Inhabitat <http://inhabitat.com/solar-powered-smart-palms-in-dubai-offer-beach-goers-wifi-and-charging-stations/solar-smart-palm>, 2015.

[10] R. Komp, S. Susan Kinne, C. Orr, Laminating PV Modules With Eva Using Solar Ovens. <http://www.mainesolar.org/EVA.pdf>.

[11] R. Komp, J. Burke, S. Kinne, M.A. Lopez, M. Lopez, J. Noel, et al., Simplified method of encapsulating fragile PV cells for cottage industry production of photovoltaic modules, in Solar 2010 Conference Proceedings, American Solar Energy Society, 2010.

[12] R. Komp, Assembling your own solar modules, in Practical Photovoltaics: Electricity from Solar Cells, third ed. rev, Aatec publications, Ann Arbor, Michigan, 2001.

[13] D.C. Miller, J.H. Wohlgemuth, Examination of a Junction-Box Adhesion Test for Use in Photovoltaic Module Qualification, Presented at SPIE Optics + Photonics 2012 San Diego, CA, 2012.

[14] NREL, NREL Research Identifies Increased Potential for Perovskites as a Material for Solar Cells. <http://www.nrel.gov/news/press/2015/21588>, 2015.

[15] B. O'Regan, M. Gratzel, A low-cost, high-efficiency solar cell based on dye sensitized colloidal TiO_2 films, Nature 353 (1991) 737–740. <http://www.nature.com/nature/journal/v353/n6346/abs/353737a0.html>.

[16] N. Papageorgiou, Dye-sensitized solar cells rival conventional cell efficiency. Ecole Polytechnique Federale De Lausanne (EPFL) News Mediacom. <http://actu.epfl.ch/news/dye-sensitized-solar-cells-rival-conventional-ce-2/>, 2013.

[17] K. Pfluke, Photovoltaic Module Assembly Using SMT Materials and Processes. Renewable Energy World. Originally published in June 2009. <www.indium.com/solar>, 2013.

[18] S. Qazi, F. Qazi, Nanotechnology for Photovoltaic Energy: Challenges and Potentials, Handbook of Research on Solar Energy Systems and Technologies, IGI Global, Hershey, PA, 2013.

[19] Reports of International Work. Maine Solar Energy Association. <http://www.mainesolar.org/Komp.html>.

[20] D.E. Regt, J.D. Diaz, S. Pai, Solar cell phone charger. D-Lab Final project report. <http://ocw.mit.edu/courses/edgerton-center/ec-711-d-lab-energy-spring2011/projects/MITEC_711S11_proj_rptchrg.pdf>, 2011.

[21] RETScreen International <http://www.retscreen.net/>.

[22] S. Sakets, Smart solar palm trees power Wi-Fi, phones in Dubai. Thomas Reuters Foundation. Edited by Jumana Farouky. <http://www.reuters.com/article/solar-dubai-palms-idUSL-5N1122XH20150827>, 2015.

[23] Z. Shahan, Phone & Tablet Chargers In NYC—Street Change, Clean Technica 2013. <http://cleantechnica.com/2013/06/19/public-solar-powered-cell-phone-tablet-chargers-in-nyc-street-change/>.

[24] SolarGIS iMaps. <http://solargis.info/doc/about-imaps>.

[25] SolarGIS pvPlanner. <http://solargis.info/pvplanner/#tl=Google:hybrid&bm=satellite>.

[26] V.D. Stranks, J.H. Snaith, Perovskite Solar Cells Could Beat the Efficiency of Silicon. Scientific American. <http://www.scientificamerican.com/article/perovskite-solar-cells-could-beat-the-efficiency-of-silicon/>, 2015.

[27] Surface Meteorology and Solar Energy (SSE) Release 6.0 Methodology Version 3.1.2 May 6, 2014. <https://eosweb.larc.nasa.gov/sse/documents/SSE6Methodology.pdf>.

[28] M. Šúri, T. Cebecauer, A. Skoczek, SolarGIS: solar data and online applications for pv planning and performance assessment, in 26th European Photovoltaics Solar Energy Conference, Hamburg, Germany, 2011.

[29] Sustainable Development in Nicaragua Grupo Fenix. <http://grupofenix.org/>.

[30] M. William, S. Wilcox, Solar Radiation Data Manual for Flat-plate and Concentrating Collectors. NREL/TP-463-5607, National Renewable Energy Laboratory, 1617 Cole, Boulevard, Golden, CO 80401, 1994.

[31] WRIGHTGRID: Secure, Connect, Empower. "Secure Charging Station." <http://www.wrightgrid.com/solutions/#.VoBgavkrJgY>.

[32] Y. Yang, D.P. Ostrowski, R.M. France, K. Zhu, J.V.D. Lagemaat, J.M. Luther, et al., Observation of a hot-phonon bottleneck k in lead-iodide perovskites, Nat. Photon. (2015).

[33] L. Zimm, Belgrade Students Design Solar-Powered Cell Phone Charging Station. Under Environment, Green Technology, Innovation, Renewable Energy, Solar Power. <http://inhabitat.com/belgrade-students-design-solar-powered-cell-phone-charging-station/>, 2011.

Index

W

Water as heat transfer fluids, 207
Water Pasteurization Indicator (WAPI), 163
Water purification, 153–162
  direct solar water treatment for. *See* Direct solar water treatment for water purification
  mobile PV systems for, 88–91
  reverse osmosis water purification, standalone PV system for, 157–160, 159*f*
    example of, 158–160
  solar suitcase for, 121–122, 122*f*
  standalone PV systems for ultrafiltration, 160–162, 161*f*
    example of, 161–162
  ultraviolet sterilization, standalone PV systems for, 154–157
    example of, 155–157
Water supply to homes or villages, standalone PV system for, 150–151, 151*f*
WCS (We Care Solar) suitcase (yellow) for medical relief in remote areas, case study, 130–136

Weather related power outages
  causes of
    cascading effect, 8
    electric lightning, 8
    rain and flooding, 8
    snow and ice, 8
    wind, tornadoes, and hurricanes, 8
  cost of, 9
Weather-related disasters, 4
Wi-Fi hotspots, 170
Wind, tornadoes, and hurricanes
  and power outage, 8
World Metrological Organization (WMO), 39
World Radiation Data Center (WRDC), 39
World Solar Challenge, 98
World solar energy map, 38*f*
WrightGrid Model Z public phone charging system, 259, 260*f*
WSS (We Share Solar) suitcase (blue) for lighting in schools and orphanages, case study, 131–136

Printed in the United States
By Bookmasters